Collins

Teacher Pack

AF754795

EDEXCEL INTERNATIONAL GCSE (9-1) PHYSICS

Steve Bibby and Malcolm Bradley

HarperCollins Publishers
Since 1817

William Collins' dream of knowledge for all began with the publication of his first book in 1819. A self-educated mill worker, he not only enriched millions of lives, but also founded a flourishing publishing house. Today, staying true to this spirit, Collins books are packed with inspiration, innovation and practical expertise. They place you at the centre of a world of possibility and give you exactly what you need to explore it.

Collins. Freedom to teach

Published by Collins
The News Building
1 London Bridge Street
London
SE1 9GF

Browse the complete Collins catalogue at:
www.collins.co.uk

10 9 8 7 6 5 4 3 2

ISBN: 978-0-00-823623-6

British Library Cataloguing in Publication Data

A Catalogue record for this publication is available from the British Library

Original material by Malcolm Bradley
New material by Steve Bibby
Commissioning Editor: Joanna Ramsay
Development Editor: Tim Jackson
Project manager: Maheswari Pon Saravanan
Project editor: Vicki Litherland
Formatting: Jouve India
Copy editor: Gwynneth Drabble
Proofreader: Mitch Fitton
Answer checker: Peter Batty
Cover design: Ink-tank
Production: Rachel Weaver
Cover photo: DrHitch/Shutterstock
Printed and bound by CPI

Health and Safety: All practicals in this book have been safety checked and the publisher and authors have made every reasonable effort to ensure that the experiments and activities in this book are safe when conducted as instructed. However, the publishers assume no responsibility for any damage or injury caused or sustained while performing the experiments or activities in this book to the full extent permitted by law.

Parents, guardians, and/or teachers should supervise students who undertake the experiments and activities in this book, and carry out full risk assessments before carrying out any practical work.

Acknowledgements
Every effort has been made to trace copyright holders and to obtain their permission for the use of copyright materials. The publishers will gladly receive any information enabling them to rectify any error or omission at the first opportunity.

Contents

Introduction

Welcome to the Collins Edexcel International GCSE Physics Teacher Pack, which has been written by experienced authors and teachers and fully safety checked to help you deliver an effective and successful International GCSE Physics course.

Overview

The **printed book** mirrors the Student Book topics and includes:

- **Activity plans** with resource lists, learning outcomes and objectives, and detailed guidance.
- **Practical activities** including teacher demonstrations and student experiments.
- **Worksheets** for students with instructions to follow and questions to answer.
- **Technician's notes** so practicals can be safely planned and executed.
- **Answers** to all the questions in the Student Book.
- **Scheme of work** that maps all practical activities and shows at a glance all the topics linked with their learning outcomes and Edexcel specification references.
- Matching chart to the **Double Award** specification.

Note that each of the topics in this Teacher Pack start with b) as the first topic. The Student Book shows the Units students will learn but this section has been omitted from the Teacher Pack.

Activity plans

The Teacher Pack has been matched topic by topic to the Student Book to allow easy cross referencing between the books. Each topic is introduced with a brief introduction, links to other topics and an overview of the activities in an easy-to-navigate table. Activity plans give guidance on delivering learning episodes that can be combined for longer lessons to give you complete flexibility and control.

Activity plans cover the following:

- **Learning objectives and outcomes** linked to the specification and the Student Book.
- **Common misconceptions** that students hold which may need addressing.
- **Resources** you will need for activities.
- A **detailed pick-up-and-teach approach** for differentiated, varied tasks.
- **Demonstration and practical activity** technician's notes so you can easily check you have everything you need.
- **Answers** to all the questions in the Student Book.

Activity plans are supported by **Worksheets, Technician's notes** and **Answers**.

Worksheets

Worksheets provide engaging activities for students. Those that deal with practical work give step-by-step guidance for students to ensure that they carry out their investigations safely and successfully.

Technician's notes

Detailed **Technician's notes** are provided for every lesson with lists of resources needed for practical activities and/or demonstrations, set-up instructions and safety notes.

For Today's Safety Information, which will need to be checked before setting up any practical activity or demonstration, please refer to your employer's guidelines or visit: www.science.cleapss.org.uk

Answers

Every activity plan in this book includes the **answers** to the associated questions in the Student Book. Answers to the **Exam-style questions** are given at the end of every section.

Scheme of work

The detailed **Scheme of work** at the back of this book provides a comprehensive overview of the Edexcel International GCSE course and the Level 1 and 2 Edexcel Certificates, matching each lesson to the learning outcomes in the specification. It also shows at a glance where the practical activities and demonstrations occur so that you can plan ahead.

Introduction

The specification content is divided into eight sections, each containing a number of different topics:

1 Forces and motion

2 Electricity

3 Waves

4 Human influences on the environment

5 Solids, liquids and gases

6 Magnetism and electromagnetism

7 Radioactivity and particles

8 Astrophysics

When you are determining the teaching sequence for the course, you should keep in mind these important considerations:

- There should be a logical progression in the development of concepts and knowledge in order to aid students' understanding.
- The course should have a 'storyline' so that wherever possible the sequence can be easily explained to students.
- Consider carefully how to introduce the more difficult concepts. For example, ensure that they are not introduced too early in the course, which may demotivate students, or put in a sequence where they all come one after another.
- Maintain the emphasis on practical and investigative work throughout the course.
- You will need to allocate different amounts of teaching time to different topics, depending on levels of difficulty or extent of the content.
- It is important to retain some flexibility so that you can revisit difficult ideas or apply new ideas in different contexts.

Suggested teaching sequence

The sequence is mapped over five terms on the assumption that much of the time in the sixth term will be allocated to revision. Where the course is started early or where the time allocated each week is either generous or limited, you will need to make adjustments.

First year, 1st term

1b Movement and position

2b Mains electricity

4b Energy

6b Magnetism

First year, 2nd term

3b Properties of waves

5b Density and pressure

2c Energy and voltage in circuits

1c Forces, movement, shape and momentum (1)

First year, 3rd term

1c Forces, movement, shape and momentum (2)

7b Radioactivity

4c Work and power

6c Electromagnetism

Second year, 1st term

5c Change of state

2d Electric charge

6d Electromagnetic induction

4d Energy resources and electricity generation

Second year, 2nd term

7c Fission and fusion

3c The electromagnetic spectrum 3d Light and sound

5d Ideal gas molecules

8b Motion in the Universe

8c Stellar evolution

8d Cosmology

Notes

- Topic 1c *Forces, movement, shape and momentum* has been divided. This is to spread out a substantial amount of content.
- Topic 4d is fairly short.

Matching chart to Double and Single Awards

This chart matches the statements in the Edexcel Double and Single Awards to the content of the Student Book.

The numbers in the columns correspond to the statement numbers in the specification. A **bold** number with a 'P' reference relates to content that is only in the IGCSE in Physics.

Section 1 Forces and motion

Topic	Physics	Double Award	Single Award	Pages
a) Units	1.1	1.1	1.1	10
	1.2P			10
b) Movement and position	1.3	1.3	1.3	12
	1.4	1.4	1.4	15
	1.5	1.5	1.5	17
	1.6	1.6	1.6	19
	1.7	1.7	1.7	21
	1.8	1.8	1.8	21
	1.9	1.9	1.9	22
	1.10	1.10		20
c) Forces, movement, shape and momentum	1.11	1.11	1.11	30
	1.12	1.12	1.12	31
	1.13	1.13		33
	1.14	1.14		33
	1.15	1.15		33
	1.16	1.16	1.16	34
	1.17	1.17	1.17	36
	1.18	1.18	1.18	38
	1.19	1.19	1.19	42
	1.20	1.20	1.20	43
	1.21	1.21		41
	1.22	1.22		55
	1.23	1.23		51
	1.24	1.24		51
	1.25P			56
	1.26P			56
	1.27P			56
	1.28P			57
	1.29P			61
	1.30P			44
	1.31P			45
	1.32P			47
	1.33P			48

Section 2 Electricity

Topic	Physics	Double Award	Single Award	Pages
a) Units	2.1	2.1	2.1	78
b) Mains electricity	2.2	2.2		79
	2.3	2.3		82
	2.4	2.4	2.4	83
	2.5	2.5		86
	2.6	2.6	2.6	87
c) Energy and potential difference in circuits	2.7	2.7		93
	2.8	2.8	2.8	95

Topic	Physics	Double Award	Single Award	Pages
	2.9	2.9	2.9	96
	2.10	2.10	2.10	97
	2.11	2.11		99
	2.12	2.12	2.12	
	2.13	2.13	2.13	101
	2.14	2.14	2.14	104
	2.15	2.15		105
	2.16	2.16	2.16	106
	2.17	2.17		105
	2.18	2.18		
	2.19	2.19	2.19	103
	2.20	2.20		108
	2.21	2.21		108
d) Electric charge	**2.22P**			117
	2.23P			120
	2.24P			119
	2.25P			120
	2.26P			121
	2.27P			122
	2.28P			122

Section 3 Waves

Topic	Physics	Double Award	Single Award	Pages
a) Units	3.1	3.1	3.1	136
b) Properties of waves	3.2	3.2		137
	3.3	3.3	3.3	138
	3.4	3.4	3.4	139
	3.5	3.5	3.5	140
	3.6	3.6		141
	3.7	3.7	3.7	140
	3.8	3.8		142
	3.9	3.9	3.9	141
c) The electromagnetic spectrum	3.10	3.10	3.10	148
	3.11	3.11	3.11	149
	3.12	3.12	3.12	151
	3.13	3.13	3.13	151
d) Light and sound	3.14	3.14	3.14	159
	3.15	3.15	3.15	160
	3.16	3.16		160
	3.17	3.17	3.17	163
	3.18	3.18		163
	3.19	3.19		166
	3.20	3.20	3.20	167
	3.21	3.21	3.21	167
	3.22	3.22		167
	3.23	3.23	3.23	170
	3.24P			171
	3.25P			173
	3.26P			174
	3.27P			175
	3.28P			176
	3.29P			176

Section 4 Energy resources and energy transfers

Topic	Physics	Double Award	Single Award	Pages
a) Units	4.1	4.1	4.1	188
b) Energy	4.2	4.2	4.2	189
	4.3	4.3	4.3	193
	4.4	4.4	4.4	194
	4.5	4.5	4.5	195
	4.6	4.6		198
	4.7	4.7		200
	4.8	4.8		200
	4.9	4.9		204
	4.10	4.10		205
c) Work and power	4.11	4.11	4.11	212
	4.12	4.12	4.12	212
	4.13	4.13	4.13	215
	4.14	4.14	4.14	215
	4.15	4.15	4.15	216
	4.16	4.16	4.16	218
	4.17	4.17	4.17	219
d) Human influences on the environment	**4.18P**			227
	4.19P			234

Section 5 Solids, liquids and gases

Topic	Physics	Double Award	Single Award	Pages
a) Units	5.1	5.1	5.1	246
b) Density and pressure	5.2	5.2	5.2	246
	5.3	5.3		247
	5.4	5.4		248
	5.5	5.5	5.5	254
	5.6	5.6	5.6	256
	5.7	5.7		258
c) Change of state	**5.8P**			266
	5.9P			266
	5.10P			267
	5.11P			269
	5.12P			271
	5.13P			272
	5.14P			273
d) Ideal gas molecules	5.15	5.15	5.15	278
	5.16	5.16	5.16	280
	5.17	5.17	5.17	280
	5.18	5.18	5.18	281
	5.19	5.19	5.19	281
	5.20	5.20	5.20	282
	5.21	5.21		282
	5.22	5.22		284

Section 6 Magnetism and electromagnetism

Topic	Physics	Double Award	Single Award	Pages
a) Units	6.1	6.1	6.1	294
b) Magnetism	6.2	6.2		296
	6.3	6.3		296
	6.4	6.4	6.4	297
	6.5	6.5		300
	6.6	6.6	6.6	301
	6.7	6.7	6.7	303
c) Electromagnetism	6.8	6.8	6.8	306
	6.9P			306
	6.10P			307
	6.11P			311
	6.12	6.12	6.12	311
	6.13	6.13	6.13	313
	6.14	6.14	6.14	315
d) Electromagnetic induction	6.15	6.15		321
	6.16	6.16		321
	6.17P			324
	6.18P			326
	6.19P			324
	6.20P			327

Section 7 Radioactivity and particles

Topic	Physics	Double Award	Single Award	Pages
a) Units	7.1	7.1	7.1	338
b) Radioactivity	7.2	7.2	7.2	340
	7.3	7.3	7.3	340
	7.4	7.4	7.4	343
	7.5	7.5	7.5	344
	7.6	7.6	7.6	349
	7.7	7.7		345
	7.8	7.8		347
	7.9	7.9		348
	7.10	7.10	7.10	348
	7.11	7.11		349
	7.12	7.12	7.12	350
	7.13	7.13		350
	7.14	7.14	7.14	352
	7.15	7.15	7.15	354
	7.16	7.16	7.16	354
c) Fission and fusion	7.17	7.17	7.17	362
	7.18	7.18	7.18	363
	7.19	7.19	7.19	363
	7.20	7.20		364
	7.21	7.21		365
	7.22	7.22	7.22	363
	7.23	7.23		362
	7.24	7.24		367
	7.25	7.25	7.25	367
	7.26	7.26		367

Section 8 Astrophysics

Topic	Physics	Double Award	Single Award	Pages
a) Units	8.1	8.1	8.1	376
b) Motion in the Universe	8.2	8.2	8.2	378
	8.3	8.3	8.3	383
	8.4	8.4	8.4	383
	8.5	8.5	8.5	381
	8.6	8.6		382
c) Stellar evolution	8.7	8.7	8.7	387
	8.8	8.8	8.8	387
	8.9	8.9	8.9	388
	8.10	8.10		389
	8.11P			388
	8.12P			390
d) Cosmology	**8.13P**			394
	8.14P			394, 398
	8.15P			394
	8.16P			395
	8.17P			397
	8.18P			398

Section 1. Forces and motion

Contents

Overview of the section

In this section students are introduced to the physics of motion in the first topic. They learn how to plot and interpret distance–time and velocity–time graphs. They learn and use the relationships between average speed, distance and time, and acceleration, velocity and time. They learn how to determine distance travelled from the area under a velocity–time graph. Experiments in the first topic investigate the motion of everyday objects such as toy cars or tennis balls.

The second topic contains many different aspects of forces. Newton's laws are covered, as is momentum, and how safety features in cars are designed in order to reduce the force experienced by the passenger in a collision. The second topic also covers centre of gravity, and moments, and Hooke's law.

Starting points

The Student Book provides a 'section opener', a double-page spread which sets the scene for students (see pages 8–9). It is structured in this way:

- an overview providing details of the areas of study
- six questions, the first two relating to aspects of physics that should be reasonably familiar to most students, and the final four relating directly to some of the topics to be studied
- a list of section contents.

The questions provide a structure for introducing the section.

1. What happens when the driving force on a car is greater than the friction force on the car? Why?
2. An animal has a mass of 28 kg. If the animal were able to travel to the Moon, would it weigh more or less there than it does on Earth? Give a reason for your answer.
3. If an object is stationary, what must be true about the forces acting on it?
4. In physics, what do we mean when we say an object is accelerating?
5. What is the force that keeps all the bodies in the Universe moving as they do?
6. A person is standing on the floor. What forces are acting on the person? Is the person creating a force on anything?

The section opener has two main purposes:

- to acknowledge the student's prior learning and to value it
- to provide a benchmark against which future learning can be compared.

The six questions or 'starting points' in the Student Book can be used in a number of different ways to introduce the section:

- You could ask students to consider the questions as an introductory homework task.
- You could put students into groups to share their own ideas and understanding and then report back to the whole class.
- Students could be given access to the Internet, preferably with a tight timescale, to find out the information required.

You could then use a spider chart or other form of wall chart to summarise everybody's ideas. Recording these initial ideas in this way allows you to retain them for reference as the individual topics are developed. In this way, your students' progress in learning can be readily acknowledged.

P1b Movement and position

Introduction

This topic introduces students to the physics of motion. It includes graphical representation of motion and how to interpret it. Students learn the definition of acceleration and the equation used to calculate it.

Links to other topics

Section	Essential background knowledge	Useful links
1 Forces and motion		P1c Forces, movement, shape and momentum
4 Energy resources and energy transfers		P4c Work and power

Topic overview

P1b.1	**Distance–time graphs, speed and velocity** This activity introduces the concept of speed. It explains the difference between average and instantaneous speed and the measurement system and units used. Students learn how to describe motion in a graphical form, and recognise key aspects of graphs of motion and relate these to the motion itself.
P1b.2	**Acceleration and velocity–time graphs** In this activity students are introduced to velocity as a vector quantity, shown how to define acceleration as rate of change of velocity and how to make measurements of accelerated motion. Students describe accelerated motion in graphical form and learn to recognise key features of such graphs. They link velocity–time graphs to acceleration and distance travelled.
P1b.3	**Consolidation and summary** This activity provides an opportunity for a quick recap on the ideas encountered in the topic as well as time for the students to answer the End of topic questions in the Student Book.

Activity P1b.1 Distance–time graphs, speed and velocity

Specification reference: P1b.1.3, P1b.1.4, P1b.1.5

Learning objectives

- Use measurement systems and the appropriate units
- Describe motion in a graphical form by plotting and explaining distance–time graphs
- Use the relationship between average speed, distance moved and time taken:

$$\text{average speed} = \frac{\text{distance moved}}{\text{time taken}}$$

Learning outcomes

- Explain the importance of having a consistent measuring system
- Draw graphs of distance against time for data given in different forms
- Plot and explain distance–time graphs (approach, recession, stationary)
- Be able to explain the difference between average speed and speed at a particular moment
- Describe average speed as the distance travelled in a set time
- Be able to investigate the motion of everyday objects, such as toy cars or tennis balls

Common misconceptions

Less able students may need to be trained to follow the line of a distance–time graph from left to right in real time, being aware of the variable being plotted on the vertical axis.

Some students think that speed and velocity are the same thing. Make sure that they are aware that speed does not have a direction associated with it but velocity does.

Students need to get used to the idea that it is up to them to define which direction is positive for velocity. Although it is standard practice to have up as positive, right could just as easily be positive as negative.

Resources

Student Book pages 10–18

Worksheet P1b.1a Measuring

Worksheet P1b.1b Distance–time data

Worksheet P1b.1c Distance–time graphs

Resources for class practicals and demonstration (see Technician's notes, following)

Resources for the introduction: Approx. 6 pictures of things that move, for example, vehicles, animals, planets, plants.

Approach

Introduce the activity

Ask the students what they understand by the word 'speed'. Where do they come across this concept in everyday life? Draw a large set of labelled distance–time axes on the board. Don't bother to put scales on the axes. Ask a student to describe their journey from home to school that morning. Sketch a graph on the axes to summarise the journey.

Give students a set of pictures (printed or on screen) and asking them to organise in order from the fastest to the slowest. Have a range, including vehicles, animals, planets, plants, etc. Depending on the students you are dealing with, choose a list where the speed in question may be a maximum speed, average speed, speed of part of the object (e.g. an engine) and so on.

The purpose is to generate comments that will lead to discussion that defines speed – both average and instantaneous. Emphasise that although there are a number of units for speed commonly in use, in physics we will stick to m/s where possible.

Develop the activity

Students measure a distance and a time. Worksheet P1b.1a gives instructions on measuring a distance and a time.

Alternatively, you could use ticker tape and ticker timers, with students pulling the tape through by hand. If you use this method, ask students to vary how the tape is pulled through – steady speed, getting faster, slowing down.

Show students how to organise the tape into a graph by taking groups of 5 or 10 dots at a time. Alternatively, if you have light gates available, make use of them to measure a simple motion such as a ball rolling across a table.

Discuss average speed with students. All that is known is the total distance travelled and the total time taken. If you have multiple lights gates you can improve on this method.

An alternative is to use simple stop clocks – or the timing function on many watches or phones, to time a simple motion which could be students walking across a classroom. If possible, extend the activity so that running, etc. is possible BUT MAKE SURE RISK ASSESSMENTS ARE CARRIED OUT AND THAT NO STUDENT IS EXPOSED TO ANY KIND OF RISK.

Move on to discuss how graphs can be used to describe motion. The practical on Worksheet P1b.1b could be carried out at this point. Students will line up along a long corridor, playing field, quadrangle or hall.

Each pair of students will have a stop clock and will work out their distance from the start of the line. At a signal from the teacher, all students start their clocks together.

They note the times at which the teacher is in front of them.

The teacher should start from 0 m and walk–run–stand along the line for about a minute.

After the practical is complete, a computer and data projector will simplify the business of sharing data so that students can draw the distance–time graph. Students can plot the points for their own graph as they are entered – they do not need to wait until all of the data has been entered.

Students can attempt Worksheet P1b.1c after they have completed the practical activity. Note that the tasks on the Worksheet become progressively more demanding. In question 2, students may need some guidance about how to calculate the speed from the graph.

As a class, draw up a checklist for interpreting distance–time graphs – constant speed looks like ..., etc. Check that key ideas are secure by providing data (from the end of topic or exam-style questions) and asking them to plot the graphs and interpret the shapes obtained. This is important as velocity–time graphs (covered in the next activity) have similar shapes but carry different meanings.

Extend students' thinking by showing them a pendulum swinging. Ask them to sketch the distance–time graph for this motion. If you have a motion sensor linked to a data logger, draw the graph out in real time to check their ideas.

Students can tackle the Developing investigative skills feature on page 13 of the Student Book in pairs. Discussing the questions will help students to clarify and develop their understanding.

Bring the activity together

Sketch a distance–time graph on the board and challenge a student to act out the journey that it represents. Allow their peers to criticise and/or approve the student's interpretation. Repeat as time permits.

Technician's notes

Be sure to check the latest safety notes on these resources before proceeding.

The following resources are needed for the class demonstration, P1b.1:

pendulum
motion sensor and data logger (optional)

The following resources are needed for class practical P1b.1a, per group.

30 cm ruler to measure length to nearest mm
stopwatch
measuring cylinders containing different volumes of liquid (enough for one for each group of students)

The following resources are needed for class practical P1b.1b, per group:

stop clock
metre rule
pad and pencil

Answers

Worksheet P1b.1c

1.

a)

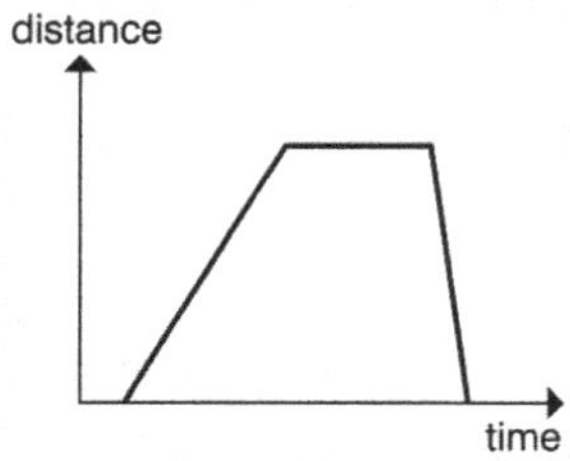

b)

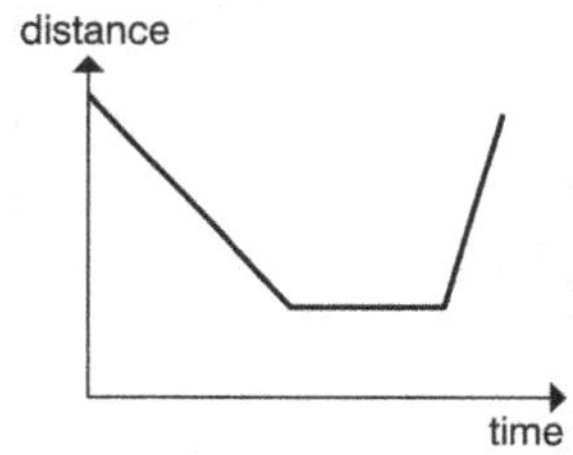

c)

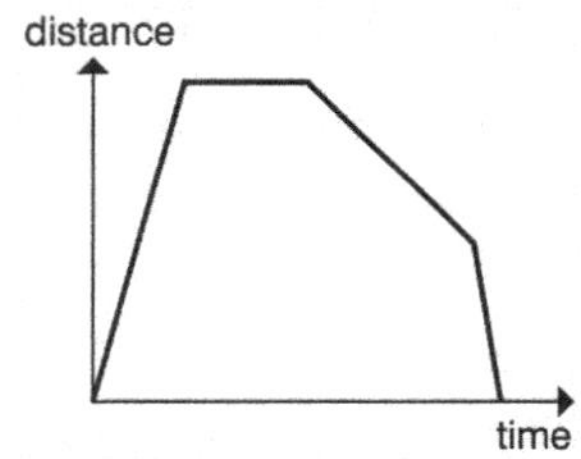

2. a) 80 km/h.
   b) 65 km/h.
   c) 130 km/h.
   d) 0 km/h.
   e) 60 km/h.

Page 11

1. Length, time, mass.
2. So that the unit remains constant and science remains comparable.
3. a) Base: kilogram, second, metre.
   b) Derived: metre/second, newton.

Page 15

1. Stationary – the distance does not change as time increases.

2. A straight line – positive gradient for moving away, negative gradient for moving towards.

3. Student's own graph – likely to be a curve, as shown in Fig. 1.2.

Page 16

1. 10 000/(15 × 60) = 10 000/900 = 11.1 m/s.

2. 22.5 m.

3. 3000 s = 50 minutes.

Page 17

1. Speed describes how far an object travels in a particular time. Velocity is numerically identical to speed, but it also includes a statement of the direction of the travel.

2. The sign of a velocity indicates a direction. 'Positive' and 'negative' velocities are in opposite directions to each other. We still need to explain which direction we are taking to be positive – this will change depending on the motion we are studying. For example, if we are looking at the motion of objects acting under the force of gravity, we might say upwards is the positive direction, so that downwards becomes the negative direction.

3. a) Diagram showing tennis ball moving upwards with a speed of +5 m/s.
   b) Diagram showing tennis ball moving downwards with a speed of –5 m/s.

Developing investigative skills, pages 17–18

1. Extra friction provided by pulling the tape along / through the ticker timer would reduce the acceleration (and the trolley may reach a constant speed if the drag was enough to balance the force pulling the trolley down the slope).

2. An alternative approach could be to video the trolley rolling down the ramp and then measuring the position frame by frame on playback. This would need a scale to be attached alongside the ramp for measurements. Using light-gates would be more difficult, since a lot would be needed if the motion of the trolley was to be studied all the way down the ramp.

3. Table completed.

4. Graph starts at the origin and curves upwards, indicating an increase in speed as the trolley moved down the ramp.

5. Table completed.

6. Graph starts at the origin and provides a (reasonably) straight line indicating a (reasonably) steady increase in speed.

7. The data point for 0.6 s looks to be incorrectly calculated as it is significantly above the pattern of the other points on the speed–time graph.

8. If yes – would allow averages to be taken, ironing out some of the random friction effects of pulling the tape through. If no – not possible to repeat exactly as the friction from the tape / path of the trolley would be slightly different each time.

P1b.1a Measuring

In this practical you will practise using different measuring instruments.

Apparatus

ruler (30 cm marked in mm)

stopwatch

measuring cylinder (marked in cm^3)

SAFETY INFORMATION

Ensure that you do not have any health issues which may prevent you taking part in physical activity.

Method

Using a ruler

For this activity you should work with a partner.

1. Draw four straight lines on a piece of paper.

2. Ask your partner to measure the lines to the nearest millimetre.

3. Record the measurements that your partner makes in a table like the one on the right.

4. Now measure the lines yourself. Record the measurements in a table like the one on the right.

Line	Length (mm)
Line 1	
Line 2	
Line 3	
Line 4	

Using a stopwatch

For this activity you should work in a small group.

1. Decide who is going to use the stopwatch first.

2. The person with the stopwatch should time how long it takes each person in the group to jump on the spot 10 times. Record the times in a table like this.

3. Repeat step 2 until each person in the group has used the stopwatch.

Name	Time to jump on spot 10 times (s)

Using a measuring cylinder

For this activity you should work in a small group.

1. Pour some water into a measuring cylinder.

2. Take turns to read the volume of water in the measuring cylinder.

Record your results in a table like the one on the right.

Name	Volume of water (cm^3)

Analyse and interpret data

1. Do any of the measurements you and your partner made in the ruler activity differ? If they do, discuss why this may have happened with your partner.

2. In the timing activity, why might the times recorded for each person be different with a different person using the stopwatch?

Evaluate data and methods

1. What do you need to do to make an accurate measurement of a length?

2. If you are timing something, what should you do to get the most precise measurement?

3. When using a measuring cylinder, how do you obtain an accurate reading?

P1b.1b Distance–time data

In this practical you will investigate the relationship between distance and time and record your findings.

Apparatus

stop clock

metre rule

notepad

pencil

Method

1. Stand in a line with your partner.
2. Find out what number you are in the line. The first person is 0, the next is 1 and so on.
3. Use the metre rule to make sure that you are 2.0 m away from the people on either side.
4. Work out your distance from the start of the line.
5. When the teacher gives the sign, start your stop clock.
6. Dictate the time to the nearest second to your partner each time that the teacher is in front of you.
7. Share your data with the rest of the group.

Analyse and interpret data

1. Use the data to draw a distance–time graph for the teacher.
2. Use the graph to write about the teacher's journey.

P1b.1c Distance–time graphs

1. Here are some descriptions of journeys. Sketch a distance–time graph for each journey. Assume that the distance is measured from the person who doesn't move.

a) Alice stands next to Bella for a while. She then walks away from Bella pauses and then runs back again.

b) Charlie starts off a long way from Dave. He walks towards Dave. When he gets halfway, he stops and then runs back to where he started.

c) Erica starts off next to Fiona. She runs away immediately. After a short stop to catch her breath, she slowly walks back, but halfway there runs back to Fiona.

2. Here is the distance–time graph for a car journey.

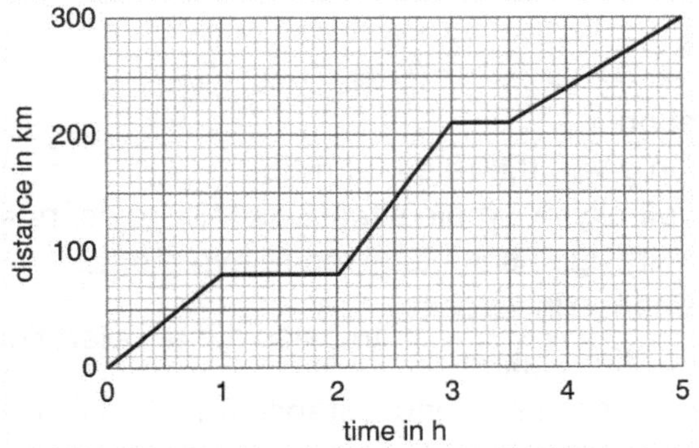

Use it to calculate:

a) the speed for the first hour

b) the speed for the last hour

c) the greatest speed

d) the lowest speed

e) the average speed of the car for the whole journey.

Activity P1b.2 Acceleration and velocity–time graphs

Specification reference: P1b.1.6, P1b.1.7, P1b.1.8, P1b.1.9, P1b.1.10

Learning objectives

- Introduce velocity as a vector quantity
- Define acceleration as rate of change of velocity
- Use the relationship between acceleration, change in velocity and time taken:

  $\text{acceleration} = \frac{\text{change in velocity}}{\text{time taken}}$, $a = \frac{(v - u)}{t}$
- Plot and explain velocity–time graphs, determining acceleration from the gradient and distance travelled from the area between the graph and the time axis
- Carry out calculations using the equation $v^2 = u^2 + (2 \times a \times s)$

Learning outcomes

- Know the definition of acceleration
- Be able to calculate acceleration given the relevant data
- Describe how to measure the velocity in simple situations (e.g. toy cars accelerating across a table top)
- Be able to draw a graph of velocity against time, given relevant data
- Be able to use $\text{acceleration} = \frac{\text{change in velocity}}{\text{time taken}}$ to calculate acceleration
- Be able to determine acceleration (gradient) and distance travelled (area under graph) from a velocity–time graph
- Be able to calculate one of final speed, initial speed, acceleration and distance moved, using the equation $v^2 = u^2 + (2 \times a \times s)$

Common misconceptions

This may be a student's first introduction to a vector quantity, one which has both a value and a direction, so it is worth making sure that they understand that the positive direction always has to be defined for vectors in one dimension.

Note that students will have a varied experience of squared numbers and square roots. This is a useful opportunity to reinforce these mathematical skills. A particularly use is that you can use this idea as a check on a calculation. Given that you cannot take a square root of a negative number then if you think you need to, you've made a mistake somewhere in your calculation. This can encourage students to be critical of their approach and therefore retrace their steps if they need to.

Students may need reminding that the beam of infrared light passing through light gate is invisible.

Resources

Student Book pages 19–25

Worksheet P1b.2a Measuring acceleration

Worksheet P1b.2b Acceleration calculations

Resources for class practical and demonstration (see Technician's notes, following)

Approach

Introduce the activity

Talk about a car's speedometer reading 50 km/h and then a few seconds later it displays 70 km/h, and explain that this means the car has accelerated.

Demonstrate the use of a light gate to measure the speed of a small object, such as a book. Ask the students to calculate the speed of the book from its length and the time recorded by the software. If a data projector is used, you could also take this opportunity to teach the class how to use the software. Don't show them how to use the software to calculate speed.

Hold a white golf ball in the air. Release it. Ask the students to describe its motion as it falls, bounces and rises back towards your hand. Help them along with the ideas of acceleration and positive and negative velocities.

Develop the activity

Discuss 'Calculating acceleration' on pages 19 and 20 of the Student Book. Students should attempt the worked examples without the given solutions to start with, and then their solutions can be compared to the given solutions and differences discussed. The questions on page 20 give further practice.

Students can then attempt the Worksheet P1b.2a in small groups. Make sure that students are aware of the safety note at the end of the sheet. The motion should involve an acceleration so that the graph is not horizontal.

The next part of the activity focuses on velocity–time graphs. Discuss the ideas on pages 21–22 of the Student Book before students attempt the questions on page 22. Make sure that students understand how to find distance from a velocity–time graph and then ask students to complete Worksheet P1b.2b. The worksheet contains additional questions on calculating acceleration and velocity–time graphs.

Students can tackle the questions 1–4 on page 24 in pairs and more able students can work on question 5, also in pairs.

Bring the activity together

Release a golf ball towards the floor.

Ask students to sketch velocity–time graphs for the ball as it falls. Once they have all managed that, you could show them again and ask them to draw an acceleration–time graph (constant negative value with a positive spike where it hits the floor, if up is positive).

Run a short test to see if students can recall the meaning of the symbols u, v, a, t and s as well as their units and the equations involving acceleration.

Technician's notes

Be sure to check the latest safety notes on these resources before proceeding.

The following resources are needed for the class demonstration, P1b.2:

light gate and data logger
data projector (if available)
golf ball

The following resources are needed for class practical P1b.2a, per group:

track
block about 5 cm high to raise it at one end
trolley
sticky tape
ruler
stop clock
light gate, clamp and stand, square of card and data logger OR ticker timer and ticker tape

Answers

Worksheet P1b.2b

1. a) +5 m/s^2. b) –4 m/s^2. c) –5 m/s^2. d) +10 m/s^2.

2. a) 75 m.
   b) +0.5 m/s^2.
   c) 268.75 m.
   d) Back the way she came. e) 187.5 m.
   f) 0.5 s.
   g) –10 m/s.

3. a) $v^2 = u^2 + (2 \times a \times s)$
   $v^2 = 0^2 + (2 \times 10 \times 159)$
   $v^2 = 3180$
   $v = \sqrt{3180}$
   $v = 56.39$ m/s
   b) i) $v^2 = u^2 + (2 \times a \times s)$
   $(55.5)^2 = 0^2 + (2 \times 10 \times s)$
   $3080.25 = 20 \times s$
   $s = \frac{3080.25}{20}$
   drop height (at high tide), $s = 154.01$
   ii) tide height = drop height at low tide − drop height at high tide
   tide height = 159.00 − 154.01
   tide height = 4.99 m (3 s.f. is OK)

Page 20 (top)

1. Acceleration = (change in velocity)/ (time taken) = rate of change of velocity.
2. Deceleration.
3. 10 m/s^2.
4. –15 m/s^2.

Page 20 (bottom)

$v^2 = u^2 + (2 \times a \times s)$
$v^2 = 0^2 + (2 \times 1.6 \times 1.5)$
$v^2 = 0 + 4.8$
$v = \sqrt{4.8}$
$v = 2.12$ m/s (to 2 s.f.)

Page 22

1. Constant velocity or speed in a straight line.

2. The line will be a straight line (constant gradient) – positive gradient for speeding up, negative gradient for slowing down.

3. Acceleration.

Pages 24–25

1. The acceleration is the gradient of the graph.

2. The distance travelled is the area under the graph.

3. a) Athlete A: 8 m/s, athlete B: 6.25 m/s.

   b) Athlete A: horizontal line at 8 m/s, starting at 0 s and finishing at 50 s. Athlete B: horizontal line at 6.25 m/s, starting at 0 s and finishing at 64 s.

   c) Area under speed–time graph is the same, i.e. 8 m/s × 50 s = 6.25 m/s × 64 s.

4. 45 m.

5. a) 10 m/s.

   b) The object is travelling at a steady speed.

   c) C and D.

   d) It is decelerating.

   e) 120 m.

   f) 10 m/s^2.

P1b.2a Measuring acceleration

You are going to find the speed of a moving object and then calculate its acceleration.

Apparatus

track
block
trolley
ruler
light gate, computer, card, clamp and sticky tape (OR ticker tape and ticker timer)
stop clock

SAFETY INFORMATION

If this experiment is carried out on a bench or desk, beware of the trolley running off the end of the slope onto the floor or someone's foot.

Method

1. Put the track on the bench with the block under one end so that it slopes gently.
2. Put the trolley on the track and fasten the length of card to the top.
3. Clamp the light gate near the lower end of the track, such that the card breaks the infrared light beam when the trolley passes by.
4. Set up the software so that it measures the time it takes the card to pass through the light gate. Measure the length of the card.
5. Put the trolley at the top of the track. Release it, starting the stop clock at the same time.
6. Stop the clock when the card is halfway through the light gate. Write the value for the acceleration time into the results table below.
7. Repeat steps 5 to 6 three times.

Analyse and interpret data

Length of card (m)	Light gate time (s)	Final velocity (m/s)	Acceleration time (s)	Acceleration (m/s^2)
			Average value	

1. Use the time for the light gate and the length of the card to calculate a value for the final velocity *v* of the trolley.
2. The initial velocity *u* of the trolley is zero. Calculate the acceleration *a* of the trolley from the values of *u, v* and *t* (the stop clock reading).
3. Repeat for all your results.
4. Calculate an average value for the acceleration.

P1b.2b Acceleration calculations

1. All of these questions assume that objects moving to the right have a positive velocity. Calculate the average acceleration of each of these objects. Don't forget to include its sign (+ or –) as well as its value.

   a) A car moves to the right from rest, reaching a speed of 40 m/s in a time of 8 s.

   b) An athlete starts from rest and after 2.5 s is moving to the left with a speed of 10 m/s.

   c) An aeroplane touches the landing strip at a speed of 60 m/s to the right. It comes to a stop 12 s later.

   d) A snooker ball rolls to the left at a speed of 5 m/s, to be hit by the table edge for a time of 0.2 s, leaving it rolling to the right at a speed of 3 m/s.

2. The graph is for a motorcyclist on a long straight track. She starts moving at 7.5 m/s. Use data from the graph to answer these questions.

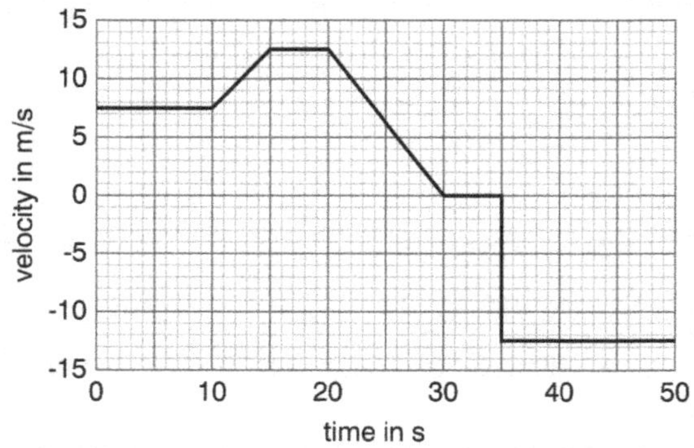

   a) How far did the motorcyclist move between 0 s and 10 s?

   b) What was her acceleration between 10 s and 20 s?

   c) How far did she move altogether before she stopped briefly at 30 s?

   d) Which direction was she moving from 35 s onwards?

   e) How far was she from the starting point at 50 s?

   f) If her acceleration at 35 s was –25 m/s^2, for how long did she accelerate?

   g) At 50 s she applies the brakes for 0.5 s. The brakes provide an acceleration of –5 m/s^2. What is her velocity when the brakes are released?

3. A large stone is dropped from a cliff. From rest it falls 159 m before it hits the sea at low tide.

   a) Calculate the stone's velocity as it hits the sea. Take the gravitational field strength as 10 N/kg.

   b) At high tide the level of the sea rises. Another large stone is dropped from the cliff. It too falls from rest but hits the sea with a velocity of 55.5 m/s.

      i) Calculate how far the stone falls.

      ii) Calculate the height of the tide.

Activity P1b.3 Consolidation and summary

Learning objectives

- To review the learning points of the topic
- To test understanding through answering questions

Learning outcomes

- Be familiar with the knowledge and understanding summarised in the End of topic checklist
- Be able to apply this knowledge and understanding by answering End of topic questions

Resources

Student Book pages 26–28

Approach

Introduce the activity

Ask students to work with a partner to make a list of key words from this topic. They could then work together to produce a spider diagram showing how the different concepts are linked. They could compare their list with the list of key words given on page 26 of the Student Book. Discuss the checklist given on page 26 and use questioning to see how much of the content students are comfortable with.

Develop the activity

Ask students to work individually to work through the End of topic questions on pages 27–28 of the Student Book without reference to the text. As they work, walk around the classroom observing their answers and questioning them as necessary to find out which questions are causing difficulties.

Bring the activity together

After a set period, ask the students to stop working and discuss any areas of difficulty you observed as you walked round the class. Students should complete any unanswered questions for homework, but you should stress that they should attempt the questions without reference to the text to see how much content they have retained.

Answers

End of topic questions mark scheme

Question	Correct answer	Marks
1	0.36 km/minute	1
2	5 m/s	1
3	–6 m/s +4 m/s	1 1
4	1.25 m/s	1
5 a)	10 km/h/s	1
5 b)	2.78 m/s^2	1
6 a)	the cars move along the road in the same direction, a constant distance apart	1
6 b)	if they are moving towards each other, they will collide head on. If they are moving apart, they will continue to do so	1
6 c)	the cars will move along the road, a constant distance apart in the opposite direction to (a)	1
7	zero you had a positive velocity in on your way to school and then a negative velocity on your way back from school. The two cancel out	1 1
8 a)	10 m/s	1
8 b)	160 m	1
8 c)	7.2 m/s work out the total distance by adding the 200 m and the 160 m together. Work out the total time by adding up the time for each part of the graph – don't forget to include the time waiting at the junction	1
9	a graph will have axes which are labelled properly with a scale, plotted data points and the best line a sketch graph is drawn to illustrate the main shape of the line only key values such as where the line changes direction given on the axes	1 1
10	using the usual way of labelling the axes, moving away has a positive gradient and moving towards you a negative gradient	1 1
11 a)	the speed remained constant if the line on a distance–time graph has a constant gradient then the speed is constant	1

Question	Correct answer	Marks
11 b)	600 m this can be read off the graph: after 20 s the car had travelled 200 m, after 60 s it had travelled 800 m. So the distance travelled is 830 – 250 = 580 m	1
11 c)	15 m/s – answer must include units speed = distance/time, v = 580/40 = 14.5 m/s < 15 m/s	1
11 d)	it stopped the line is horizontal during this time, indicating that the car was not moving (no distance was travelled)	1
12 a)	2.0 m/s^2	1
12 b)	½ x (0.75 × 1.5) + ½ x (2.25 × 1.5) 2.25 m	1 1
	Total:	26

P1c Forces, movement, shape and momentum

Introduction

In this topic students are introduced to the types of forces and the idea of a resultant force. They are introduced to Newton's laws, the concept of weight and the concept of momentum. Students look at safety in vehicles and how the relationship between force, change in momentum and time taken is used to design safety features. The topic concludes with a study of moments, stability of objects and a look at Hooke's law.

Links to other topics

Section	Essential background knowledge	Useful links
1 Forces and motion	P1b Movement and position	
4 Energy resources and energy transfers		P4c Work and power
5 Solids, liquids and gases		P5b Density and pressure
6 Magnetism and electromagnetism		P6b Magnetism
8 Astrophysics		P8b Motion in the Universe

Topic overview

P1c.1	**Introducing forces** In this activity students are introduced to the effects of forces. They learn about the types of force and about friction. The activity ends by considering how to find resultant forces.
P1c.2	**Balanced and unbalanced forces** In this activity students learn what happens when balanced forces act on an object and what happens when the forces on an object are not balanced.
P1c.3	**Force, mass and acceleration** In this activity students learn about the relationship between force, mass and acceleration. It includes a formal statement of Newton's first and second laws of motion and the application of the laws in a variety of situations.
P1c.4	**Mass and weight** In this activity students are introduced to the concepts of mass and weight and the difference between them.
P1c.5	**Falling objects** In this activity students learn about falling objects. There are two suggested practicals on terminal velocity in this activity.

P1c.6	**Factors affecting stopping distances** This activity describes how the stopping of a vehicle is related to factors involving (a) the driver, (b) the vehicle and (c) the external conditions. Students learn how to describe qualitatively the effect of changing any of these factors.
P1c.7	**Turning effect of a force and stability of an object** In this activity students learn about the turning effect of a force. They investigate the centre of mass, centre of gravity and how the stability of an object depends on the position of the centre of gravity. Finally they learn about the principle of moments and the conditions for equilibrium.
P1c.8	**Stretching** In this activity students learn about force–extension graphs and Hooke's law. Students learn about the elastic behaviour of a spring. Students also learn about non-linear stretching and the behaviour of rubber.
P1c.9	**Momentum** In this activity students are given a definition of momentum and introduced to the principle of conservation of momentum. The students are shown applications of Newton's second law to simple situations.
P1c.10	**Safety features in vehicles** Students are introduced to seatbelts, airbags, crumple zones, etc. as examples of safety features designed into vehicles. These are linked to the rate of change of momentum.
P1c.11	**The third law** This activity introduces Newton's third law and the idea of a third law pair.
P1c.12	**Consolidation and summary** This activity provides an opportunity for a quick recap on the ideas encountered in the topic as well as time for the students to answer the End of topic questions in the Student Book.

Activity P1c.1 Introducing forces

Learning objectives

- Describe typical effects of forces – changing the speed, shape or direction of objects
- Identify different types of force
- Understand that force is a vector quantity
- Calculate the resultant force in simple situations (one dimension only)
- Know that friction is a force that opposes motion

Learning outcomes

- Be able to name at least two different types of force
- Be able to draw and interpret diagrams involving multiple forces in familiar situations

Common misconceptions

The idea that if forces in one direction are positive, then forces in the opposite direction are negative does not come easily to many students. To get good results in the practical work requires skill and patience. Lots of repeats should iron out a lot of the random errors, but won't help with systematic ones such as incorrect zeroes on the force meters.

Resources

Student Book pages 29–34

Worksheet P1c.1a Finding forces

Worksheet P1c.1b Multiple forces

Resources for class practical and demonstration (see Technician's notes, following)

Approach

Introduce the activity

Stretch a long length of rubber tubing to remind students of the effect of forces on materials. Suggest to them that the length of the rubber tubing tells you about the size of the force stretching it. Introduce a force meter as a device which uses the length of a spring to measure a force. Show them how to set the zero and take readings from it.

Show students a video clip of a drag racer accelerating and then slowing down using a parachute – challenge the students to explain in detail why the car accelerates and then stops.

Roll a ball or toy car across a table – again challenge the students to explain the motion in detail. The idea of a number of forces all acting at the same time on an object is critical to students' understanding of the topic from now on so draw upon as many everyday examples as you or the students can come up with.

Develop the activity

Discuss the effects of forces. Use Student Book page 30 to assist with the discussion. Talk about the various types of force. Students should tackle the questions on page 31 of the Student Book in pairs. You can then discuss the Science in context feature on page 32 on linking the fundamental forces.

Ask students to give you examples of vector quantities that they have already met (velocity). Ask students to work in pairs to make a table listing scalar and vector quantities. Take feedback.

Discuss 'Scalars and vectors' on page 33 of the Student Book. Students should answer the questions on page 34 in pairs. Discuss responses.

Take the example of the cars given on page 33 of the Student Book. With the active involvement of students, work through the diagrams so that the students understand the resultant force in each case. Give the students several more examples until they are confident with the techniques involved. (You may find it useful to discuss this with your mathematics colleagues who may be able to provide you with suitable examples for the students to work on.)

Discuss the nature of friction and how it is sometime useful and sometimes something which is a disadvantage. At this point students can carry out the practical activity. Details are given on Worksheet P1c.1a. If you do not have sufficient equipment, this could be done as a demonstration but make sure that students are fully involved by using questioning and involving students in doing the activity.

The practical gives students some practice in thinking about some of the forces acting on an object on a ramp (friction, weight and pull). The investigation can expand or contract to fit the time available. Students may be delighted to find that the friction force is proportional to the mass of the object when the ramp is horizontal. There is no such easy pattern as the slope of the ramp is increased; the pulling force must become the weight of the sled when the ramp is vertical.

Students can now tackle the questions on Worksheet P1c.1b before discussing the questions on page 34 of the Student Book in pairs.

Bring the activity together

Discuss the results of the practical. Ask students to tell you two things that they thought worked well and one thing that they would change if they had a chance to repeat the practical.

Ask students to work with a partner to make a summary of the main points of the activity.

Technician's notes

Be sure to check the latest safety notes on these resources before proceeding.

The following resources are needed for the class demonstration, P1c.1:

rubber tubing
force meter
video clip of a drag racer accelerating and slowing down using a parachute
ball or toy car

The following resources are needed for class practical P1c.1a, per group:

force meter
ramp
block to raise one end of ramp
sled
slotted masses
ruler
string loop
balance

A sled which has a raised lip at the back will stop masses sliding off when the ramp is steep.

Beware of trolley falling off the end of the slope onto feet. A cardboard box or piece of soft material on the floor will catch the trolley/soften its landing. Or put a stop on the end of the ramp, e.g. a pile of textbooks.

Answers

Worksheet P1c.1b

1. a) 130 N forwards.

   b) 90 N to the left.

   c) 6 N downwards.

   d) 0 N.

2. a) i) B ii) C iii) A

   b) Zero.

   c) Although A and B could balance, there needs to be another upwards force to balance out C which acts downwards.

Page 31

1. Effects – forces can change the speed of an object, the shape of an object, the direction the object is moving in. Types of force – gravitational, electric, magnetic, (electromagnetic), strong nuclear force.
2. Gravity.
3. The mass of objects and the distance between their centres.
4. Electromagnetic force.

Page 34 (top)

1. Vector has size and direction, scalar just has size.
2. a) resultant force = 1000 − 100 = 900 N forward

   b) resultant force = 1000 − 500 = 500 N forward

   c) resultant force = 1000 − 1000 = 0 N

   d) resultant force = 1000 − 1200 = −200 N forward (or 200 N backwards)
3. 2560 N.

Page 34 (bottom)

1. Walking, driving – if there is no friction, you skid.
2. Where energy transferred to thermal energy which is lost to the surroundings.

P1c.1a Finding forces

You are going to investigate the factors which affect the force required to pull an object up a ramp.

Apparatus

force meter

ramp

block

sled

slotted masses

ruler

string loop

balance

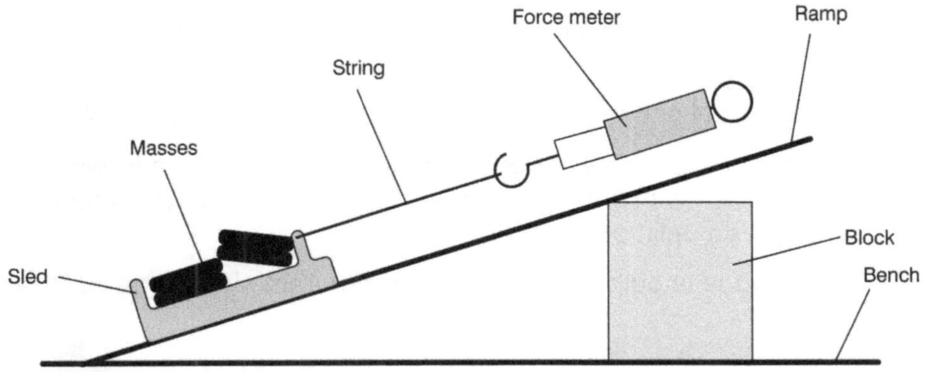

a stop at the end of the ramp would be advisable, e.g. a pile of textbooks, or a box on the floor to catch any falling objects

Method

1. Use a block to set up the ramp on the bench as shown above.
2. Measure the distance of the top of the ramp above the bench.
3. Add some slotted masses to the sled. Use a balance to find the total mass of the sled.
4. Use the force meter to pull the sled along the ramp at a slow steady speed. Note the required force measured by the force meter.
5. Do a series of measurements with different heights of the ramp. Keep the total mass of the sled constant.
6. Do a series of measurements with different masses for the sled. Keep the height of the ramp constant.
7. Repeat step 6 with the ramp horizontal on the bench so that the force meter is measuring just the friction on the sled.

Total mass of sled (kg)	**Height of ramp (m)**	**Force on meter (N)**

Analyse and interpret data

1. How did the force on the meter vary with the height of the ramp?
2. How did the force on the meter vary with the mass on the sled?

Evaluate data and methods

1. Suggest how you could have improved this investigation.
2. Were the results what you expected?

P1c.1b Multiple forces

1. Work out the resultant force for each of these situations. State both the value and the direction.

   a) The engine pushes a car forwards with a force of 250 N and the brakes provide a backwards force of 120 N.

   b) Two men push a piano to the left. One pushes with a force of 100 N, the other with a force of 80 N. Friction between the piano and the floor provides a force of 90 N to the right.

   c) A rock falls through water. Gravity pulls it down with a force of 42 N. The water pushes up with a force of 16 N and friction provides an upwards force of 20 N.

   d) An airplane has four engines, each of which provides a forwards thrust of 5000 N. Air resistance pushes back with a force of 20 000 N. The weight of the airplane is 60 000 N. Each of the two wings provides an uplift of 30 000 N.

2. A block of wood is pulled up a ramp. The diagram shows three of the forces acting on it.

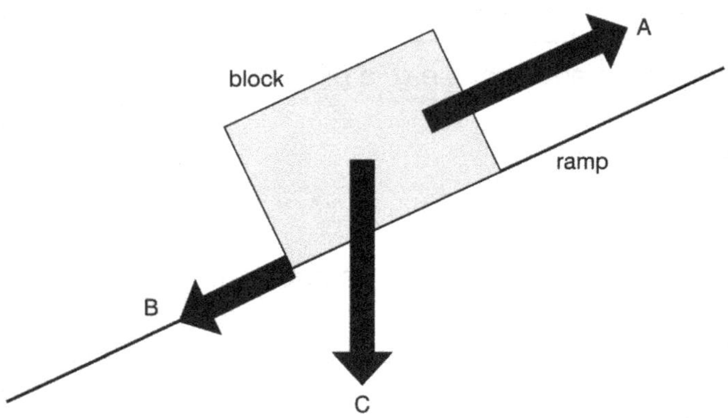

   a) Which letter shows each of the following forces?

      i) friction between the block and the ramp

      ii) force of gravity on the block

      iii) force pulling the block up the ramp.

   b) The pulling force is adjusted until the block stops moving. What is the resultant force on the block when this happens?

   c) Explain why at least one more force must be acting on the block when it is not moving.

Activity P1c.2 Balanced and unbalanced forces

Learning objectives

- Investigate balanced and unbalanced forces

Learning outcomes

- Know what balanced and unbalanced forces are

Common misconceptions

Students need to appreciate that when they talk about a force, they also need to state the object upon which it acts. The action force on one object is equal and opposite to a force on another object. This does not prevent the resultant force on any object from being other than zero.

Resources

Student Book pages 35–36

Worksheet P1c.2a Balanced forces

Resources for class demonstrations (see Technician's notes, following)

Approach

Introduce the activity

Ask a student to stand on a set of bathroom scales on a low platform. The class should agree that gravity provides a downwards force (weight) on the student. Then ask what will happen when the student steps sideways off the scales.

The student's downwards acceleration suggests that whilst standing on the scales there must be another upwards force to balance the weight. This reaction force comes from the compressed spring in the scales.

Now remove scales and convince the class that the reaction force comes from the (tiny) compression of the platform. Note that all that is needed is a safe platform which can hold a student a short distance above the ground. It needs to be wide enough to accommodate the scales.

Suspend 500 g of slotted masses from a force meter held in your hand. Ask students to explain why the masses don't accelerate towards the floor. Discuss the forces acting on the masses; the downwards action from gravity and the upwards reaction from the force meter. The upwards reaction force is measured by the force meter and because the masses aren't accelerating, the downwards action force must have the same value.

Then transfer their attention to your hand. The weight of the meter and slotted masses is the action, your upwards tug must be the equal and opposite reaction.

Develop the activity

Discuss the topic 'Balanced forces' on page 35 of the Student Book. Then discuss 'Unbalanced forces' on page 35. It is important that students get used to the idea of always mentioning the object upon which a force acts whenever they discuss forces. Forces cannot be there without objects to exert them and react to them.

Students can then tackle questions on page 36 of the Student Book before they attempt Worksheet P1c.2a.

Bring the activity together

You could sit a student on a chair on a platform and use class questioning to identify pairs of balanced forces. Do not forget to include the weight of the student and the equal and opposite reaction on the Earth.

Another option (which you should undertake with great care) is to put a student on a skateboard. Choose one who is a novice at skateboarding. Ask them to move forwards **without** touching the ground. They should find this almost impossible. Now give them a large heavy ball and ask them to try again. Get the class to explain why throwing the ball forwards allows the skateboarder to start moving backwards.

Technician's notes

Be sure to check the latest safety notes on these resources before proceeding.

The following resources are needed for the class demonstrations, P1c.2:

Introduction:

platform
bathroom scales
500 g slotted masses
force meter

Bring the activity together:

skateboard
stool for student to sit on
platform to place stool on
large heavy ball

Answers

Worksheet P1c.2a

A. 80 N to the left.

B. 150 N to the left.

C. 50 N to the left.

D. 30 N to the left.

E. 12 N downwards.

F. 50 N downwards.

Page 36

1. The object will be stationary or moving in a straight line at constant speed.
2. The object will be changing speed and/or direction.
3. His weight.

P1c.2a Balanced forces

None of these objects are moving. One of the forces acting on each object is not shown. What is its size and direction?

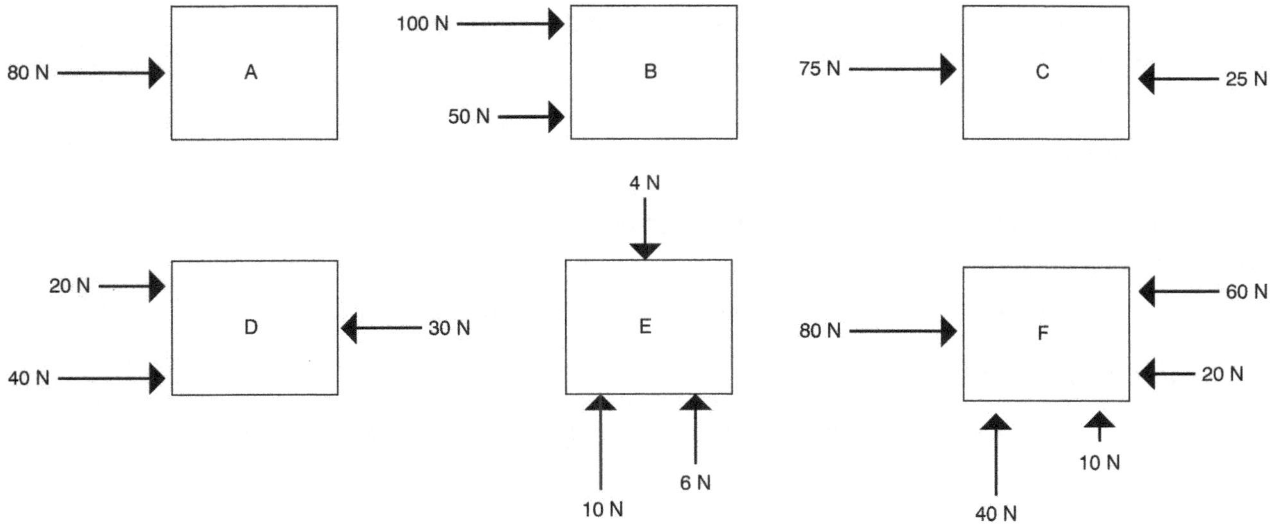

Activity P1c.3 Force, mass and acceleration

Specification reference: P1c.1.17

Learning objectives

- Know and use the relationship between force, mass and acceleration:
  force = mass × acceleration
  $F = ma$

Learning outcomes

- Understand the link between force, mass and acceleration
- Carry out calculations using $F = ma$
- Apply the principles involved in a range of familiar and unfamiliar situations

Common misconceptions

Students often confuse mass with weight. Use the word weight only for forces due to gravity.

Resources

Student Book pages 36–38

Worksheet P1c.3a Force and acceleration

Resources for demonstration and class practical (see Technician's notes, following)

Resources for the final part of the activity, per group: mini-whiteboard and marker OR large sheet of paper and marker pens.

Approach

Introduce the activity

Release a golf ball from above your head. Ask students **why** it accelerates as it moves towards the floor and decelerates on its way up again. They may suggest that there is a downward force (gravity) acting on the ball when it is moving both downwards and upwards.

Ask them to suggest what happens to the acceleration as the force on the ball and its mass are changed (but don't attempt to demonstrate it).

Develop the activity

Show students how to use a light gate and cardboard comb to measure acceleration before they tackle the practical. They will need to operate as a team in groups of four to get reasonable results, so you may prefer to run this as a whole class demonstration. Instructions are given on Worksheet P1c.3a.

There will probably not be enough time in the lesson to obtain more than one change in each value of mass and force, so agreement with the accepted formulation of Newton's second law will necessarily be rough and ready.

Discuss 'How are mass, force and acceleration related?' on pages 36–37 of the Student Book. The worked examples should be tackled by students without the given solution and then, after a period of time, students' solutions should be compared with the given solution and any differences discussed. Students can then attempt the worksheet.

You may need to remind students that a negative acceleration is a deceleration.

Students have a lot of difficulty in accepting that a resultant force of zero does not automatically result in a velocity of zero. Pointing out that in real-life scenarios, the applied forces do not usually take account of friction, so an applied force of zero will always result in slowing down.

Perceptive students may worry about the vertical forces acting on the objects considered in the practical. Reassure them that the reasons why they do not need to worry will be revealed in a later activity.

Students can then tackle the questions on page 38 of the Student Book.

Bring the activity together

Use the board to present students with a variety of forms of transport – cars, bikes, lorries, trains and planes. For each, give suggested values for mass and times to get to 100 km/h from a standing start – students might be able to help with the latter. Challenge students to rank the forms of transport by size of resultant force.

Issue whiteboards and marker pens. Set them some questions like the ones in the table below. This should allow you to assess their progress in using $F = ma$.

Question	Driving force (N)	Friction (N)	Mass (kg)	Resultant force (N)	Acceleration (m/s^2)
a)	1500		500		+2.5
b)	500	200	600		
c)	3000	500			+1.25
d)	400	1000	800		

Technician's notes

Be sure to check the latest safety notes on these resources before proceeding.

The following resources are needed for the class demonstration, P1c.3:

golf ball
hard surface on the ground

The following resources are needed for the class practical P1c.3a, per group:

dynamics trolley
track
extra mass for trolley
10 g slotted masses
pulley wheel
strong thread
light gate
cardboard comb for light gate

sticky tape to hold comb in place
ruler to measure comb teeth length
computer running appropriate software
access to scales
force meter (0–1 N)
cardboard box or similar

If you do not have light gates and data logging equipment, students can use a stop clock instead.

Students need to be aware of the hazard of the trolley running off the end of the slope and landing on their feet. A cardboard box on the floor will catch the trolley, if it falls off the bench.

Answers

Page 38

1. Force = mass × acceleration.
2. The force must be the resultant force, the mass must remain constant.
3. 600 N.
4. 800 m/s^2.

P1c.3a Force and acceleration

You are going to find out how the acceleration of an object is related to its mass and resultant force.

Apparatus

track

trolley

10 g slotted masses

pulley wheel

thread

light gate

computer

cardboard comb

scales

force meter

extra mass for trolley

sticky tape

cardboard box to catch falling masses/trolley

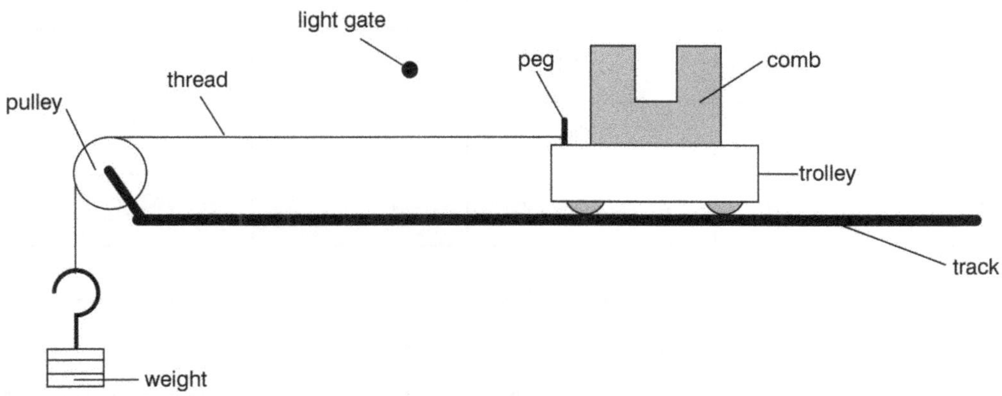

SAFETY INFORMATION

Beware of masses and/or the trolley coming off the end of the track and landing on someone's feet.

Method

1. Place the track on a level table with the pulley over the edge.
2. Fasten the cardboard comb to the top of the trolley. Find the mass of the trolley.
3. Place the trolley at one end of the track.
4. Place the light gate in front of the trolley so that the comb breaks the light beam twice on its way through.
5. Check that the computer measures the acceleration of the trolley when you accelerate it through the light gate.
6. Fasten one end of the thread to the peg at the front of the trolley.
7. Use the force meter to find the weight in N of 30 g of slotted masses. This will be the horizontal resultant force on the trolley when you release it.
8. Fasten the slotted masses to the other end of the thread, as shown in the diagram. You will need to hold the trolley in place on the track.
9. Release the trolley but catch it again before it reaches the pulley.
10. Record the acceleration, resultant force and mass of the trolley.
11. Add another 30 g of slotted masses to double the accelerating force. Repeat the experiment.

Analyse and interpret data

1. Record your results in a table like the one below.

Mass of trolley (kg)	Slotted masses (kg)	Total mass (kg)	Resultant force (N)	Acceleration (m/s^2)
	0.03			
	0.03			
	0.06			
	0.06			

2. Add the masses of the trolley and slotted masses to find the total mass which is accelerating.
3. Use the results to complete these sentences.

   Increasing only the resultant force on the trolley ________the acceleration. Increasing only the mass of the trolley __________the acceleration.
4. Here are some rules for calculating the resultant force F from the total mass m and the acceleration a.

$$F = \frac{m}{a} \qquad F = m + a \qquad F = ma \qquad F = \frac{a}{m} \qquad F = m - a$$

Which one of them is correct?

Activity P1c.4 Mass and weight

Specification reference: P1c.1.18

Learning objectives

- Know and use the relationship between weight, mass and gravitational field strength:

  weight = mass × gravitational field strength

  $W = mg$

Learning outcomes

- Know how to calculate the weight of an object using $W = mg$

Common misconceptions

The main difficulty here is that the common usage of weight is not the same as the meaning of the term in physics.

Resources

Student Book pages 38–41

Approach

Introduce the activity

Ask the students to work in pairs to discuss what they understand by the terms 'mass' and 'weight'. Take feedback. It is likely that the students will give the everyday use of weight – which is 'weighing' something and getting a quantity in grams or kilograms. This is a misconception that needs to be addressed in this activity.

Develop the lesson

Use the text on page 38 of the Student Book to introduce the meaning of the term 'mass' to describe how much material is present in a body. Go on to explain that 'weight' is the force on the object due to gravity. This concept can be reinforced by looking at the worked example on page 39 of the Student Book.

Discuss the concept of the gravitational field around the Earth. Give students some examples of masses and ask them what the weight of each mass would be on Earth.

Students should go on to compare the weight of an object on Earth with the weight of the same object on the Moon. These students should also be able to explain why it is harder to stop the supertanker in the photograph on page 40 of the Student Book than a speedboat travelling at the same speed.

Students need to discuss how objects are weighed. Use the topic 'How do you weigh something?' on page 40 of the Student Book to help with this. Students can then tackle the questions on pages 39 and 41 of the Student Book in pairs.

Finish the lesson

Ask students to work in pairs to write definitions of mass and weight that could be included in a physics dictionary. After a short time, take feedback and compare ideas. Then ask students to make a summary of the main points of the activity in a form of their choice.

Answers

Page 39

1. Mass amount of matter, weight force.
2. 600 N.
3. 50 kg.

Page 41

1. The different planets have different masses and so they have different gravitational field strengths. Since your weight = mass × gravitational field strength you will have a different weight (but the same mass) on different planets.
2. 96 N.
3. 50 kg.
4. 20 N on the left hand side.

Activity P1c.5 Falling objects

Specification reference: P1c.1.21

Learning objectives

- Describe the forces acting on falling objects
- Explain why falling objects reach a terminal velocity

Learning outcomes

- State that the acceleration of free fall for a body near to the Earth is constant
- Describe qualitatively the motion of bodies falling in a uniform gravitational field with and without air resistance (including reference to terminal velocity)

Common misconceptions

Students expect falling objects to speed up as they drop, so they always delight in the shock of coming across objects which appear to float down at a steady speed. Friction due to air resistance is a difficult concept for many to accept because air is so insubstantial.

Students also find it hard to believe that weight is independent of velocity and may cling to a belief that things reach a terminal velocity because they are moving fast enough to switch off gravity.

Resources

Student Book page 41–42

Worksheet P1c.5a Falling cupcake cases 1

Worksheet P1c.5b Falling cupcake cases 2

Resources for class practicals (see Technician's notes, following)

Resources for class demonstration: Paper cupcake cases and sheets of A4 paper (one per student)

Approach

Introduce the activity

Demonstrate the steady fall of a cupcake case. Then crumple it into a ball and let the students have a go at explaining why this makes such a difference to its fall.

Space students around the room. Give each student a piece of A4 paper. Show them how to fold it into a fan. Allow them to fan themselves vigorously. Draw their attention to the difference increase of effort required to move the fan more quickly. Introduce the idea of air resistance.

Develop the activity

Discuss the text on page 41 of the Student Book. Students can then attempt the first of the two practicals in pairs on Worksheet P1c.5a. The practicals require good hand–eye coordination on the part of the student who does the timing. The student who does the launching may want to kneel on the bench, so you need to make sure they know how to do this safely.

More able students can tackle the Extension task on page 42.

Finally, students can go through and answer the questions on page 42 of the Student Book.

Bring the activity together

If time allows students can do the practical on Worksheet P1c.5b. You could ask each group to read out their evaluation for constructive criticism by their peers.

Students should work in pairs to produce a summary of the main points of the lesson in a form of their choice.

Technician's notes

Be sure to check the latest safety notes on these resources before proceeding.

The following resources are needed for the class practical P1c.5a, per group:

one metre rule
paper cupcake case
clamp, stand and boss
stop clock (±0.01 s)

The size of the case is not very important, although larger ones may be easier for students to handle. Cases which are not unfolded by students can be reused.

The following resources are needed for practical P1c.5b, per group:

one metre rule
five paper cupcake cases
clamp, stand and boss
stop clock (±0.01 s)

Answers

Page 42 (top)

1. The velocity that an object, such as a skydiver, has when the forces are balanced so that the object travels at a maximum constant speed.

2. 10 m/s.

3. 15 N upwards.

4. 585 N downwards.

5. 9.75 m/s^2.

Page 42 (bottom)

1. a) Velocity.
 b) Force.
 c) Acceleration.

2. The velocity graph decreases from initial zero until it reaches a constant value, which is the terminal velocity. The force graph includes the constant weight (the drag), which increases with time and then becomes constant; and the resultant, which is found by adding the weight and the drag. The acceleration graph increases initially and then reaches zero because the velocity has reached terminal velocity, which is not changing, so there is no acceleration.

P1c.5a Falling cupcake cases 1

Apparatus

metre rule

clamp, stand and boss

stop clock

paper cupcake case

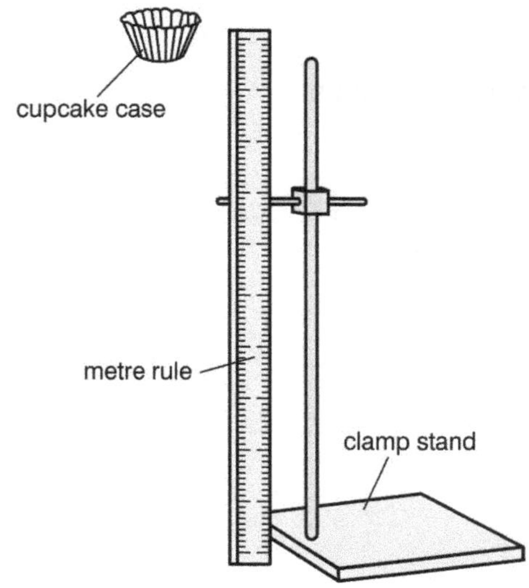

Method

1. Clamp the metre rule so that it is upright, with one end on the ground.
2. Hold the cupcake case with its base 0.50 m above the ground.
3. Release the case and start the clock at the same time.
4. Stop the clock when the case hit the ground.
5. Repeat the process twice more to check the repeatability of your results.
6. Repeat steps 3 to 5 for heights of 1.00 m, 1.50 m, 2.00 m and 2.50 m above the ground.

You will need to put the clamp stand on the bench for some of these.

7. Enter your results into a table like this.

Height (m)	Time of fall (s)			Average time (s)	Average speed (m/s)
0.50					
1.00					
1.50					
2.00					
2.50					

Analyse and interpret data

1. Use your results to plot a graph showing how the average speed depends on how long the case has been falling.

2. Try to explain the shape of the graph.

P1c.5b Falling cupcake cases 2

Apparatus

metre rule

clamp, stand and boss

stop clock

paper cupcake cases

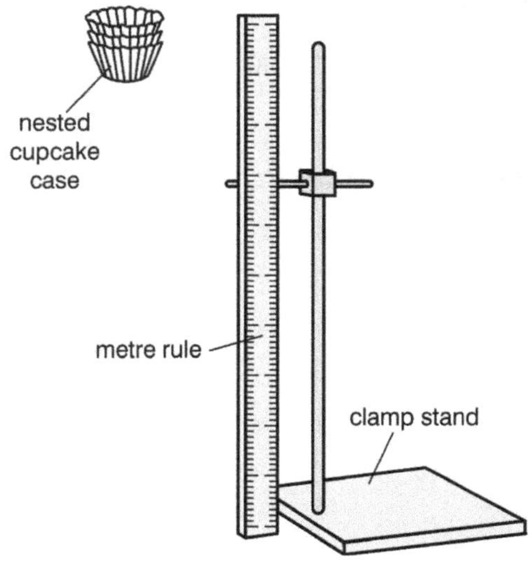

Method

1. Clamp the metre rule so that it is upright, with one end on the ground.
2. Hold the cupcake case with its base 1.00 m above the ground.
3. Release the case and start the clock at the same time.
4. Stop the clock when the case hits the ground.
5. Repeat the process twice more to establish the reliability of your results.
6. Repeat steps 2 to 5 for two, three, four and five cases nested in each other.
7. Record your results into a table like this.

Number of cases	Time of fall (s)			Average time (s)	Average speed (m/s)
1					
2					
3					
4					
5					

Analyse and interpret data

1. Use your results to plot a graph showing how the average speed depends on the number of cases.
2. Does your graph agree with your hypothesis?

Activity P1c.6 Factors affecting stopping distances

Specification reference: P1c.1.19, P1c.1.20

Learning objectives

- Know that the stopping distance of a vehicle is made up of the sum of the thinking distance and the braking distance
- Describe the factors affecting stopping distance (speed, mass, reaction time and road conditions)

Learning outcomes

- Be able to describe how the stopping of a vehicle is related to factors involving the driver, the vehicle and the external conditions
- Understand the terms thinking distance, braking distance and stopping distance
- Relate changes in these factors to an increase or a decrease in the affected distance

Common misconceptions

Students need to be persuaded that the safe distance to leave between vehicles moving along a road is much greater than the distance that they observe in everyday life.

They often ignore the special meaning of work in physics as the intermediary between energy transfers.

They are unaware that the kinetic energy of a vehicle is proportional to the square of its velocity, so have no way of appreciating why braking distance rises so alarmingly with increasing vehicle speed. Kinetic energy is covered in Topic P4c.

Be aware that there is a good chance that at least one of your students will have experienced a serious car accident. If they want to, let them share this with the class.

Students can often describe thinking **time** when they are defining thinking **distance** – watch for this.

Resources

Student Book pages 42–44

Worksheet P1c.6a Reaction time

Worksheet P1c.6b Braking distance

Resources for class practicals and demonstration (see Technician's notes, following)

Approach

Introduce the activity

Place a metal cylinder on the bench so that it rolls freely when pushed. Ask what brings the cylinder eventually to a halt. Repeat the experiment with the cylinder on its flat end so that friction is much greater. Draw out the link between braking distance and friction. Emphasise that there is a limit to the amount of friction available – hence the danger of skidding.

Show video clips of cars skidding to a stop. Give the students a list of factors that may or may not affect the distance it takes a car to stop – on cards for sorting, or on a list or using presentation software.

Develop the activity

Discuss the ideas of the students and then introduce the idea of splitting the 'stopping distance' into two parts – the thinking distance and the braking distance. Provide some government or manufacturer data and ask the students to plot a graph of speed against stopping distance.

Divide the stopping distance into the two parts and ask the students to suggest why one is a linear relationship (thinking distance is proportional to speed) whereas the braking distance is not (it depends upon kinetic energy and hence upon speed squared).

Students can then carry out the reaction time practical on Worksheet P1c.6a. Students can become quite competitive about measuring their reaction time and use up a lot of lesson time trying to improve their time – usually in vain, although there is a training effect. There are other methods of measuring reaction time, but this one neatly links it to a distance moved. The equation assumes that gravity accelerates the ruler downwards at 10 m/s^2.

You may want to use this opportunity to reinforce the message that driving with too much alcohol in your blood costs lives.

Students should then carry out the braking distance practical on Worksheet P1c.6b.

This will be the first time that students will have met the term 'work' in its special physics sense. It will be formally introduced in Topic P4c. Similarly, there is little point at this stage in establishing that the data shows that kinetic energy is proportional to the square of the speed.

More able students should tackle the Extension feature on pages 43–44 of the Student Book. All students should tackle the questions on page 43 in pairs.

Bring the activity together

Give students a couple of minutes to memorise the data for different speeds. Test them on the data. Use a data projector to show some clips of car accidents to reinforce the real-life importance of the lesson subject matter.

Technician's notes

Be sure to check the latest safety notes on these resources before proceeding.

The following resources are needed for the class demonstration P1c.6:

metal cylinder
video clips of cars skidding to a stop
cards each with a factor that may or may not affect the distance it takes a car to stop
government or manufacturer's data on stopping distances at different speeds
video clips or photos of car accidents

The following resources are needed for the class practical P1c.6a, per group:

50 cm ruler
calculator

There is a chance that the ruler will not be caught and will hit the floor or feet.

The following resources are needed for the class practical P1c.6b, per group:

light gate
computer running light gate software
metre rule
smooth bench surface
metal cylinder

The cylinder needs to be heavy and wide enough to slide stably, but not too wide so it fits easily through the jaws of the light gate.

If the surface of the bench is not smooth enough to allow the cylinder to slide without falling over, a plank of wood should be provided.

There is a risk that the cylinder will slide off the table onto the floor or feet.

Answers

Page 43

1. Thinking distance: Any one of: Speed, fatigue, use of medication/drugs/alcohol, level of distraction. Stopping distance: Speed, weather condition of road, amount of tread on tyres, friction from brakes, mass of car.

2. 36 m.

3. 18.

4. 7.5 m.

5. It would increase.

Pages 43 and 44

1. Thinking and braking distances both less.

2. Yes, reaction time is faster.

3. Velocity at 30 km/h = 8.3 m/s, at 45 km/h = 12.5 m/s. In both cases acceleration = $6.5 \times 10^3/1000 = 6.5\ \text{m/s}^2$.

Use the equation $v^2 = u^2 + (2 \times a \times s)$, which gives $u^2 = -2as$ (as $v^2 = 0$) and so $s = \dfrac{u^2}{-2a}$

At 30 km/h, $s = \dfrac{(8.3)^2}{2 \times 6.5}$ = 5.3 m

At 45 km/h, $s = \dfrac{(12.5)^2}{2 \times 6.5}$ = 12 m

P1c.6a Reaction time

This practical allows you to investigate and record your reaction time.

Apparatus

50 cm ruler

calculator

Method

1. Decide who is to be the experimenter and who is to be the subject of the experiment.
2. The experimenter holds the ruler at the 50 cm end so that it hangs down.
3. The subject places thumb and forefinger on either side of the ruler at the 0 cm end **without touching it**.
4. Without warning, the experimenter releases the ruler. The subject has to catch the falling ruler between their thumb and forefinger.
5. The subject calculates their reaction time from the distance fallen by the ruler with the help of this equation:

$$\text{reaction time in seconds} = \sqrt{\frac{\text{distance fallen in centimetres}}{500}}$$

6. The experimenter repeats steps 2 to 5 two more times so that the subject can calculate their average reaction time.
7. The subject swaps places with the experimenter and runs through steps 2 to 6.

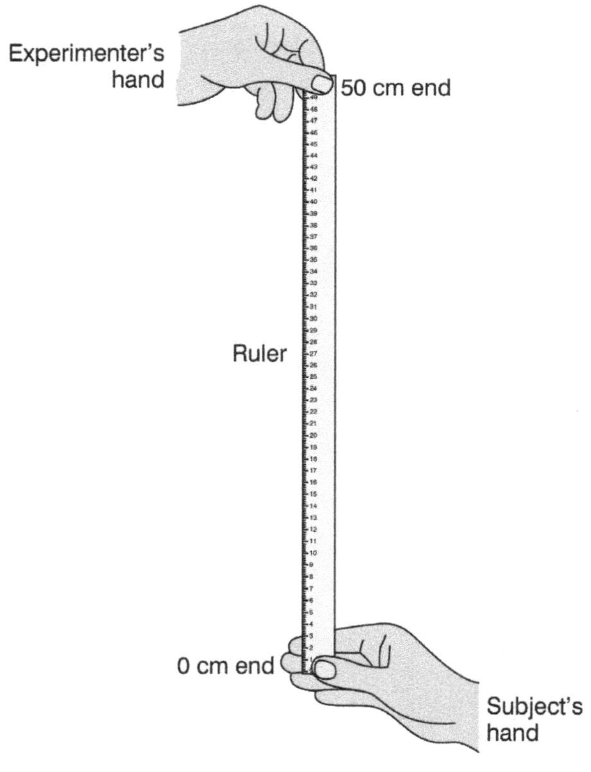

P1c.6b Braking distance

This practical allows you to investigate the relationship between initial speed and braking distance.

Apparatus

metal cylinder

metre rule

light gate

computer

bench

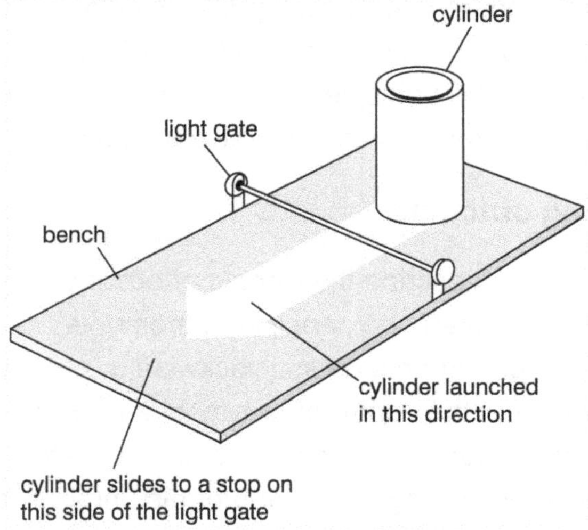

Method

1. Arrange the light gate so that the beam of infrared is just above the bench.
2. Set up the computer to measure the speed of the cylinder in m/s as it slides through the light gate.
3. Push the cylinder so that it slides freely on its flat end through the light gate.
4. Measure the distance from the light gate to the centre of the cylinder after it has come to a halt.
5. Enter the speed of the cylinder and its braking distance into a results table like the one shown below.

Initial speed (m/s)	**Braking distance (m)**

6. Repeat until you have results for a wide range of braking distances.

Analyse and interpret data

1. Use your results to plot a graph which shows how the braking distance depends on the initial speed.

Activity P1c.7 Turning effect of a force and stability of an object

Specification reference: P1c.1.30P, P1c.1.31P, P1c.1.32P, P1c.1.33P

Learning objectives

- Define moment as force × perpendicular distance from pivot to line of action of force
- Know that the weight of a body acts through its centre of gravity
- Introduce the principle of moments and conditions for equilibrium
- Balance turning forces in simple situations

Learning outcomes

- Be able to describe the turning effect of a force in simple examples
- Be able to state the principle of moments
- Describe moments as anticlockwise or clockwise and calculate moments, in N m, in simple situations
- Know what the centre of gravity of a body is
- Apply the concept of moments to explain the operation of simple devices, e.g. screwdriver
- Calculate missing forces and/or distances in situations where the forces are balanced

Common misconceptions

Most students readily accept the concept of centre of mass. However, this is often forgotten when questions are set which involve this principle, such as simple 'see-saw' moments questions where the pivot is not central.

Resources

Student Book pages 44–49

Worksheet P1c.7a Finding the centre of gravity

Worksheet P1c.7b Investigating moments

Worksheet P1c.7c Balancing forces

Resources for class practicals and demonstrations (see Technician's notes, following)

Approach

Introduce the activity

Ask students to work with a partner to make a list of things which use a force to turn something (spanner, rotary beater, etc.). Take feedback. If possible, give the students the opportunity to use the items discussed.

Discuss less obvious examples such as pushing a door handle to close a door, or a see-saw.

Develop the activity

Discuss the text on page 44 of the Student Book. Students should tackle the worked example on page 45 without the given solution initially and then, after a short period, their solution should be compared with the given solution and differences discussed. Students should then tackle the questions on page 45 in pairs.

Project a series of images of a vase, with different width bases. Students should select the vase they feel is the most stable and explain their reasons why.

Discuss the concept of centre of gravity. Use the text on page 45 of the Student Book to help. Ask students to carry out the practical on finding the centre of gravity. Instructions are given on Worksheet P1c.7a.

The practical forms a nice 'challenge' activity – the aim being for students to balance their shape perfectly on the end of a pin. Students initially find it difficult to believe this is possible, particularly for complex shapes, and are impressed when it proves possible.

Provide students with cubic and rectangular blocks (laid long end downwards). Ask them to decide which object they would define as the most stable. Discuss how stability is affected when a rectangular block is placed on one end. Explain that stability depends upon two factors; the height of the centre of mass above the base of the object and the width of the base of the object.

Students should consider the Science in context feature on page 46. They should tackle questions on page 47 in pairs.

Discuss the factors on which the moment of a force depends: the size of the force and the distance between the line of force and the pivot. Students could then attempt the worked example on page 47 as described previously.

Students can carry out the practical on Worksheet P1c.7b on moments on a beam.

Move on to discuss the conditions for equilibrium. Students can then carry out the practical on Worksheet P1c.7c on balanced forces.

Discuss 'Forces on a beam' on page 48 and then ask students to tackle the questions on page 49 in pairs.

Bring the activity together

Discuss the results of the practicals. Ask students what worked well and what one thing they would change if they repeated the practicals on another occasion.

Ask students to work with a partner to make a summary of the main points of the activity.

Technician's notes

Be sure to check the latest safety notes on these resources before proceeding.

The following resources are needed for the class demonstration P1c.7:

spanner and nut
rotary beater
model of see-saw
a series of images illustrating various degrees of stability (for example, a range of vases of differing heights and base sizes would be ideal)
computer and data projector

The following resources are needed for the class practical P1c.7a, per group:

stiff card
long pins
plumb line
clamp stand

The following resources may also be needed, per group:

two cubic blocks
two rectangular blocks

The following resources are needed for the class practical P1c.7b, per group:

balancing rulers – flat pieces of wood approximately 5 cm × 60 cm, sub-divided and marked into twelve equally sized topics, as shown below.
pivots
selection of 1 N weights
balancing ruler:

6	5	4	3	2	1	1	2	3	4	5	6

The following resources are needed for the class practical P1c.7c, per group:

two force meters
wooden bar
three loops of string
slotted masses
clamp stand and boss and heavy mass to act as a stability/counter balance for the clamp stand

Be aware that masses could fall off the mass hangers.

Students will suspend the wooden bar by a force meter at either end. Some of the slotted weights will be suspended from the bar.

Students will need to weigh the bar itself, so the string loops need to be long enough for them to make a knot.

Note: The bar can be made out of dense wood, the heavier the better. A square or round cross-section works better than a rectangular one.

Answers

Page 45

1. The size of the force and the distance between the line of the force and the turning point, which is called the pivot.

2. 2 N m.

3. 1.25 N m.

4. 0.4 m.

Page 47

1. The centre of gravity is the point where we can assume all the mass of an object is concentrated.
2. Student's own description based on method described in the text.
3. No, it will fall over.

Page 49

1. 1.5 m.
2. 20 N each.
3. F_4 = 16 N, F_3 = 24 N.

P1c.7a Finding the centre of gravity

In this practical, you will demonstrate how the centre of gravity of an irregular object can be found.

Apparatus

long pin
A4 sheet of card
scissors
plumb line
pencil
ruler
clamp stand

SAFETY INFORMATION

Take care with sharp pins.

Method

1. Draw an irregular shaped object, similar to the shape on the below. Cut it out of the card.

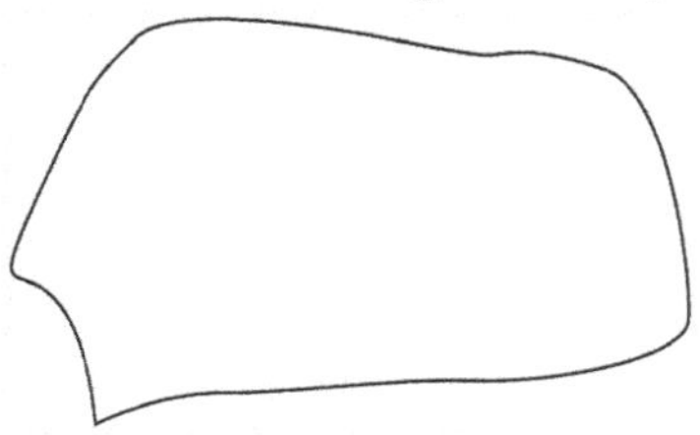

2. Insert a pin through the shape and suspend from a clamp stand. Take care when using the sharp pin.
3. Hang a plumb line from the pin; mark onto the shape the path of the plumb line.

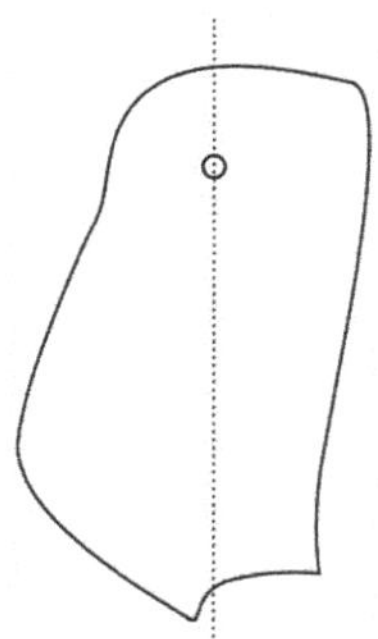

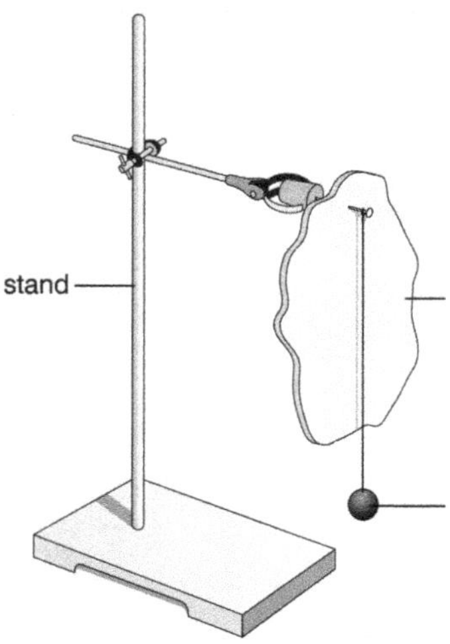

4. Remove from the clamp stand.
5. Repeat the above method for two further pin positions.
6. The point of intersection of the three lines will indicate the centre of gravity.

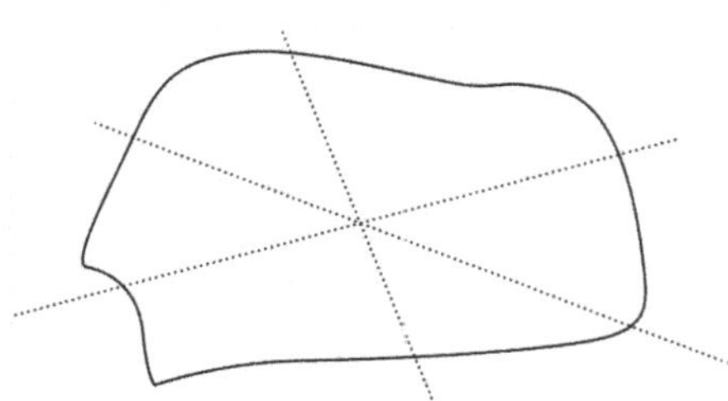

7. Check your accuracy by trying to balance your object on the end of the pin. If you have marked the centre of gravity correctly, the shape will balance.

P1c.7b Investigating moments

Apparatus

marked balancing beam pivot

selection of 1 N weights

Method

1. Set up the balancing beam on the pivot, as shown below:

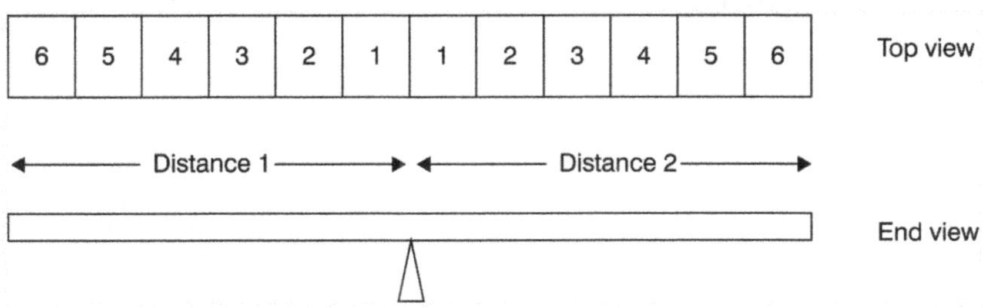

2. Place 2 N (2 × 1 N weights) at a distance of three squares left of the pivot. Take care when using masses.
3. Place 3 N weights to the right of the pivot. Slide them until the beam balances.

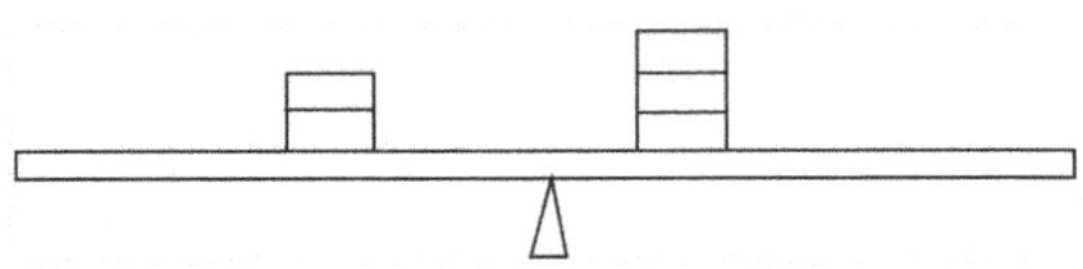

4. In the results table, note the square in which the 3 N weights are placed.
5. Repeat for all lines in the results table.
6. Complete the missing force and distance from pivot data to create a balance in the results table below:

Force 1 (N)	**Distance 1 (squares)**			**Force 2 (N)**	**Distance 2 (squares)**	
2	3			3		
4				2	4	
3	2			1		
3	4			4		
	1			2	2	
3	4				6	

Analyse and interpret data

1. Are there any patterns between the force and distance on each side of the pivot?

2. Create new headings for the blank columns: 'Moment 1' and 'Moment 2'. The units would be 'N squared' (N^2).

3. Complete the results table, calculating the moments for each line of the table.

4. Look at each line of the table in turn. What do you notice about the size of the moment to the left of the pivot (the anti-clockwise moment) and those to the right (the clockwise moment) when the beams are balanced?

5. Using the terms 'clockwise moments' and 'anti-clockwise moments', write a rule which must apply if an object is to be in equilibrium (balanced).

P1c.7c Balancing forces

You are going to do an experiment to test the hypothesis that the vertical forces on an object which is not moving add up to zero.

Apparatus

two force meters

wooden bar

three loops of string

slotted masses

clamp stand and boss

heavy mass to use as a stability/counter-balance weight on the clamp stand base

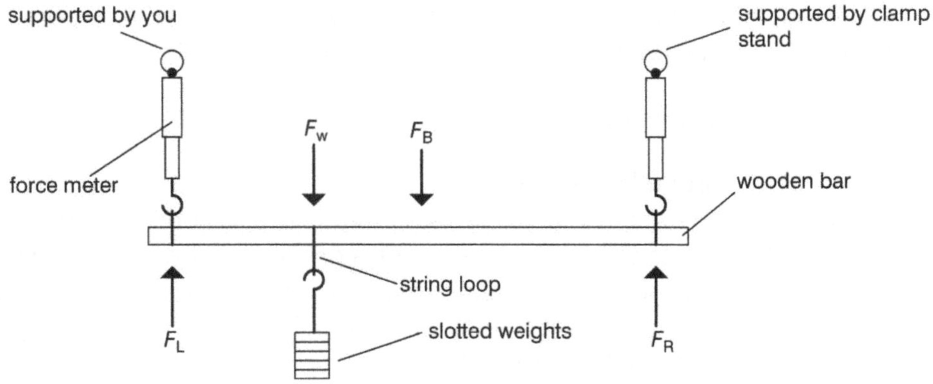

Method

1. Adjust the zero setting on both force meters.
2. Use a force meter and a string loop to find $\mathbf{F_B}$, the weight of the wooden bar. It acts downwards, so will be positive. Enter the value into the results table.
3. Hold one of the force meters above the bench with the clamp and stand.
4. Slip all three string loops onto the wooden bar.
5. Hang 5 N of the slotted masses from the middle loop.
6. Use the outer loops to suspend the bar as shown below. The slotted weights provide a downwards force $\mathbf{F_W}$ on the bar.

The force meters on the left and right provide upwards forces $\mathbf{F_L}$ and $\mathbf{F_R}$ on the bar.

7. Adjust the force meter which you are supporting until the bar is horizontal.
8. Record the readings of the two force meters. They act upwards, so will be negative.
9. Complete the calculations of the first row of the table.

The last column should be close to zero. If it isn't, go back and check all your measurements.

10. Repeat steps 5 to 9 at least five times with different numbers of slotted weights and different positions of the middle string loop.

Analyse and interpret data

1. Record your results in the table and carry out the calculation shown in the right-hand column.

$\mathbf{F_B}$	$\mathbf{F_W}$	$\mathbf{F_L}$	$\mathbf{F_R}$	$\mathbf{F_B + F_W + F_L + F_R}$
	+5.0 N			

Activity P1c.8 Stretching

Specification reference: P1c.1.22, P1c.1.23, P1c.1.24

Learning objectives

- Investigate how extension varies with applied force for springs, wire and rubber bands
- Know that the linear region of a force–extension graph is linked with Hooke's law
- Describe the elastic behaviour of a spring

Learning outcomes

- Be able to state Hooke's law
- Describe an experiment to test a spring to see if it obeys Hooke's law
- Be able to state the requirements for a graph to show direct proportionality
- Describe how not all materials follow Hooke's law when under tension
- Describe an experiment to test a material that does not give proportional behaviour (e.g. a rubber band) including safety precautions

Common misconceptions

Students are often confused by the term 'extension'. They automatically assume that it is the change of length caused by the most recent change of force, whereas it is actually the net increase from a zero-force situation. Less able students may need help in sorting out a useful scale for the axes of their graph and converting the extension into metres to calculate the spring constant.

Some students do not appreciate the use of the word 'potential' in that energy's name.

Resources

Student Book pages 49–55

Worksheet P1c.8a Stretching springs

Resources for class practicals (see Technician's notes, following)

Approach

Introduce the activity

Stretch a long piece of rubber tubing. Show students a piece of wire. Ask them whether they think it is possible to stretch the wire.

Develop the activity

Discuss 'How are materials affected by stretching?' on page 49–51 of the Student Book. Include a discussion of the Science in context feature on page 50.

Introduce Hooke's law using the text on page 51.

At this point the students should carry out the practical on Worksheet P1c.8a.

Students will suspend the hanger from the spring and measure its length for a succession of weights hung from one end.

Note that the spring will exceed its elastic limit, so cannot be reused.

After they have completed the practical, give students the following questions:

- A spring has an unstretched length of 50 mm. Its length is 65 mm when it is used to hold up a 3.0 N weight. Calculate its force constant and use it to predict the length of the spring when it holds up a 5.0 N weight.

  What do you have to assume in your calculations?

- Another spring has a length of 7.60 cm when it holds up a 4.20 N weight. The length increases to 8.20 cm when the weight is increased to 6.60 N.

Calculate the force constant of the spring and use it to find the unstretched length of the spring.

The need to convert units is a common difficulty when dealing with real-life examples where extensions are typically measured in cm or mm.

Students will also need to know how to alter the subject of an equation. This might be a good time to remind them how to do it. They also need to get used to writing down all the steps of their working as they work out the answers to the two questions above.

You can then discuss the Science in context feature on page 53. Students should look at the topic 'Equation for Hooke's law' on page 53 of the Student Book. The worked example (page 54) should be approached by asking students to come up with their own solution to the problem before comparing their solutions to the given solution and discussing any differences. The questions on page 54 provide further practice.

Ask students to study the Developing investigative skills topic on page 55. This can be done in pairs.

Bring the activity together

Set a number of simple calculations using Hooke's law to test students' understanding.

Technician's notes

Be sure to check the latest safety notes on these resources before proceeding.

The following resources are needed for the class demonstration P1c.8:

long piece of rubber tubing
piece of wire

The following resources are needed for the class practical P1c.8a, per group:

clamp stand and counter-balance weight
two clamps and bosses
disposable spring
nine 1 N weights and 1 N hanger
50 cm ruler
wooden splint
sticky putty to fasten splint to spring
eye protection

The spring will exceed its elastic limit, so cannot be reused.

Answers

Page 54

1. 0.01 N.

2. 20 N/m.

3. 20 cm.

Developing investigative skills, page 55

1. Risk of the spring unwinding at either end, causing the masses to be dropped and the spring to fly up.

2. Risk of toppling over – stand can be fixed to the table using a G-clamp. Risk of masses falling and damaging feet or floor – minimise risk by having soft material available in landing area and keep feet out of the way.

3. Preliminary experiment helps to judge the values of mass needed to achieve measurable extensions with this kind of spring. In general, preliminary ('trial') experiments give an idea of how the experiment will run so you know if it is worth the time and effort to follow the method particularly carefully.

4. Values in table, reading 1.0 N, 2.0 N, etc.

5. Values in table, reading 0 cm, 4 cm, etc.

6. Graph plotted – straight line up to the last two values where it curves away.

7. The spring obeyed Hooke's law up to a point (at a load of 12 N). This is justified by referring to the graph being a straight line (up to that point) through the origin.

8. The spring had been permanently stretched, so could not be re-used for this experiment.

9. Add a marker, such as an optical pin, to the end of the spring. This can then be set against the ruler. If the marker is not exactly at the end of the spring it does not matter, since in this experiment the student is looking for *changes* in length.

P1c.8a Stretching springs

Apparatus

metal spring
two clamps
wooden splint
eye protection
nine 1 N weights and 1 N hanger
clamp stand
50 cm ruler
sticky putty
counter-balance weight

The spring and ruler should be clamped over the base of the stand unless a heavy counter-balance weight is used to stabilise the clamp stand.

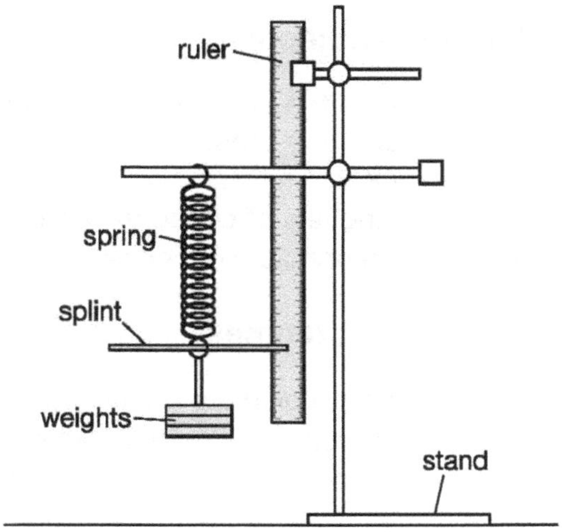

Method

1. Put on eye protection.
2. Fasten the two clamps to the stand, as shown in the diagram.
3. Clamp the ruler vertically.
4. Fasten the splint to one loop of the spring. Use sticky putty to hold it in place.
5. Hang the spring from the clamp and note the position of the splint on the ruler.
6. Hang a 1 N hanger from the end of the spring. Note the new position of the splint.
7. Add a 1 N weight to the hanger. Note the new position of the splint.
8. Remove all the weights. Check that the spring has not been permanently stretched.
9. Replace the weights and add an extra 1 N weight. Note the new position of the splint.
10. Repeat steps 8 and 9 until the spring becomes permanently stretched.

Then carry on until the total weight at the end of the spring is 10 N.

Enter your results in a table like this. Use the following equation to fill in the extension column:

Weight (N)	Splint position (mm)	Extension (mm)
0		0
1		
2		
...		
10		

extension = stretched length – unstretched length

Analyse and interpret data

1. Draw a graph to show how the stretching force depends on extension of the spring.
2. Put the label **elastic limit** on the curve at the point where the spring becomes permanently stretched.
3. Identify the straight line portion of the graph near the origin. Use it to find a value for the spring constant of the spring. Use the equation:

$$\text{spring constant (N/m)} = \frac{\text{force (N)}}{\text{extension (m)}}$$

Activity P1c.9 Momentum

Specification reference: P1c.1.25P, P1c.1.27P

Learning objectives

- Introduce the definition of momentum:
  momentum = mass × velocity
  $p = m \times v$
- Apply the law of conservation of momentum to simple situations
- Use the principle of conservation of momentum to calculate mass, velocity or momentum of objects

Learning outcomes

- Be able to calculate momentum
- Describe simple situations where momentum is conserved
- Make relevant calculations using $p = m \times v$ in simple situations

Common misconceptions

Many people appear to have an intuitive need for a quantity (often called inertia) which keeps things moving. By now, students should know that this can't be force. Momentum is a good candidate for this quantity – you can only change the motion of an object if you change its momentum, and that requires the application of a force.

Many students assume that kinetic energy is also conserved in collision or explosions. This is not usually the case, except where the two objects have the same mass (as in a two-ball Newton's cradle).

Some students have difficulty in separating scalars from vectors. Vectors terms such as high and low routinely coupled with scalar quantities, such as energy and temperature, does not help.

Resources

Student Book page 56

Worksheet P1c.9a Collisions

Worksheet P1c.9b Explosions

Resources for demonstration and class practicals (see Technician's notes, following)

Approach

Introduce the activity

Discreetly arrange for two students of obviously different masses (this is best done by choosing a tall and a short student) to stand on separate skateboards, facing each other. Ask the class to decide what will happen when the smaller person pushes the taller one away. Then ask the taller person to do the pushing instead. Ensure students are instructed to push gently.

Develop the activity

Demonstrate the equipment that they will be using the practical work. Introduce the hypothesis of momentum conservation and explain how the experiments will allow them to calculate the momentum changes in collisions or explosions.

Discuss the text on pages 56–59 of the Student Book. Be aware of the difficulties that some students have in separating scalars from vectors.

Students carry out the collisions practical on Worksheet P1c.9a.

It is important that students get accustomed to writing out their calculations line by line to show all the steps, instead of doing it in their head with the help of a calculator. The terms elastic and inelastic when applied to collisions are used somewhat loosely.

Students could then carry out the explosions practical on Worksheet P1c.9b.

Bring the activity together

Stand a light student on a skateboard. Great care will need to be taken if this activity is carried out. It may be best to choose a student who is competent on a skateboard. Get the class to predict what will happen when the student catches a heavy rubber ball that you fling at them. Then ask them to explain what will happen to the student when they fling the ball back at you. Finally, ask why they don't acquire a backwards velocity when they catch the ball (it's because friction welds you to the Earth, so your effective mass is huge).

Give students mini-whiteboards and marker pens (or students can use pen and paper). With students working in pairs, set a number of collision and explosion calculations for them to do. You could use the following questions:

- An object of mass 3.2 kg moves to the right with a speed of 5.0 m/s. It collides and sticks to a stationary object of mass 0.8 kg. Calculate the velocity of the two objects after the collision. [4.0 m/s to the right]
- An object of mass 2.4 kg moves to the right with a speed of 3.0 m/s. Another object of mass 1.6 kg moves to the left with a speed of 8.0 m/s. The objects collide and stick to each other. Calculate the final velocity of the two objects. [−1.4 m/s (to the left)]
- Jenny stands on a skateboard holding a ball. While the skateboard isn't moving, she throws the ball to the left. If she ends up moving to the right at 0.15 m/s, how fast is the ball moving when it leaves her hand?. The ball has a mass of 0.5 kg, the skateboard is 5 kg and Jenny has a mass of 55 kg. [p = 9 kg m/s, so v = 18 m/s to the right]
- An astronaut has a total suited mass of 120 kg. He is floating in space next to his spacecraft. He flings a tool to the left at 20 m/s, acquiring a speed of 0.5 m/s to the right. What is the mass of the tool? [p = 60 kg m/s, so m = 3.0 kg]

Technician's notes

Be sure to check the latest safety notes on these resources before proceeding.

The following resources are needed for the class demonstration P1c.9:

two skateboards or small trolleys
skateboard, heavy rubber ball
whiteboard and marker pen, per student OR paper and pen

You need to ensure that the parties concerned push against each other in a safe manner. The skateboards need to be oriented end to end.

The following resources are needed for the class practical P1c.9a, per group:

two 0.5 kg dynamics trolleys
sticky putty or similar if trolleys won't stick together
track
spirit level

card to fit on trolley
two light gates
computer(s) running light gate software
extra 0.5 kg masses for trolleys
access to scales (up to 1 kg)

The following resources are needed for the class practical P1c.9b, per group

two 0.5 kg dynamics trolleys, one with a plunger
wooden rod to release plunger
track
spirit level
cards to fit on both trolleys
two light gates
computer(s) running light gate software
extra 0.5 kg masses for trolleys
access to scales (up to 1 kg)

Answers

Page 56

1. 2.32 kg m/s.
2. 50 000 kg m/s.
3. 1000 kg.
4. 25 m/s.

P1c.9a Collisions

In this practical, you will investigate the effect that a collision has on the momentum of an object.

Apparatus

two dynamics trolleys
two light gates
track
trolley masses
computer
scales
spirit level
card
sticky putty

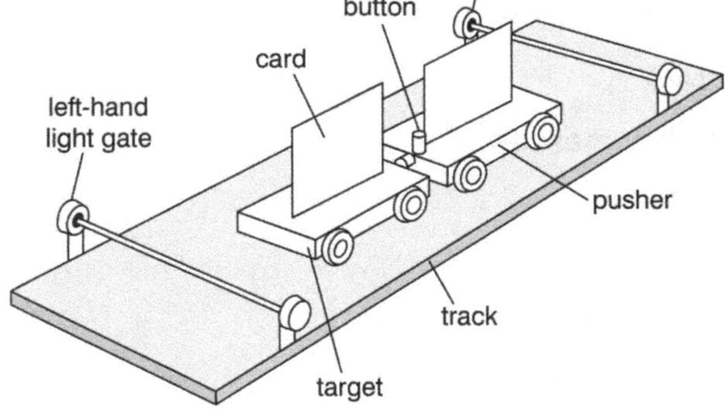

Method

1. Use the spirit level to place the track level on the bench.
2. Check that each trolley has a mass of 0.5 kg.
3. Fasten the card to one trolley. This trolley is the **bullet**. Place it at the far right.
4. Place the other trolley on the centre of the track. This trolley is the **target**.
5. Arrange matters so that the bullet and target stick together when they touch.
6. Set up the light gates so that they measure the speed of the card in m/s.
7. Place the light gates as shown in the diagram. The right-hand one measures the speed of the bullet before it hits the target. The left-hand one measures the speed of both after they have hit and stuck together.
8. Carefully launch the bullet at the target. Note the speed of the bullet before and after the collision.
9. Enter the results in a table like this and complete the calculations.

Before colliding			After colliding		
Moving mass (kg)	**Speed (m/s)**	**Momentum (kg m/s)**	**Moving mass (kg)**	**Speed (m/s)**	**Momentum (kg m/s)**
0.5			1.0		
0.5			1.5		
1.0			1.5		

10. Repeat steps 8 and 9 for the other combinations of mass of bullet and target.

Analyse and interpret data

1. What conclusion can you make about the momentum before and after each collision?

P1c.9b Explosions

In this practical, you will investigate the effect that an explosion has on the momentum of an object.

Apparatus

trolley with plunger and trolley without plunger

two cards

two light gates

track

trolley masses

computer

scales

spirit level

wooden rod

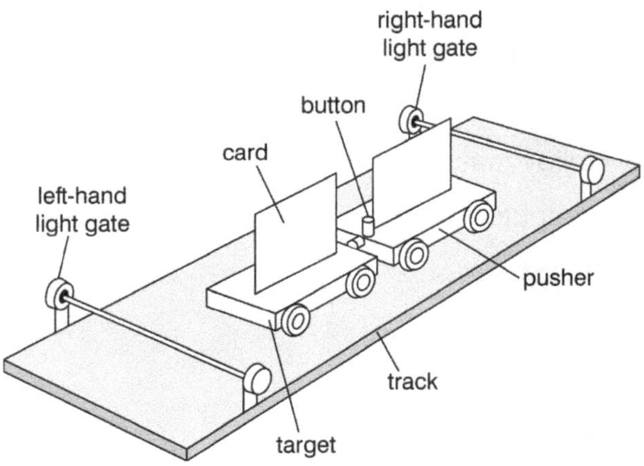

Method

1. Use the spirit level to place the track level on the bench.
2. Check that each trolley has a mass of 0.5 kg.
3. Fasten a card to each trolley.
4. Fully compress the plunger on one trolley. This is the **pusher** trolley. Place it on the centre of the track.
5. The other trolley is the **target**. Place it on the track next to the plunger.
6. Set up the light gates so that they measure the speed of the card in m/s.
7. Place the light gates as shown in the diagram. The right-hand one measures the speed of the source. The left-hand one measures the speed of the target.
8. Carefully tap the button with the wooden rod. This sets off the explosion. Note the speed of the pusher and target after the explosion.
9. Enter the speeds in a table like this and calculation the momentum for each trial.

Target			Pusher		
Mass (kg)	**Speed (m/s)**	**Momentum (kg m/s)**	**Mass (kg)**	**Speed (m/s)**	**Momentum (kg m/s)**
0.5			0.5		
0.5			1.0		
1.0			0.5		

10. Repeat steps 8 and 9 for the other combinations of mass of pusher and target.

Analyse and interpret data

1. What conclusion can you make about the momentum of the target and the pusher after each trial?

Activity P1c.10 Safety features in vehicles

Specification reference: P1c.1.26P, P1c.1.28P

Learning objectives

- Use the idea of momentum to explain safety features in vehicles, such as seatbelts, airbags and crumple zones
- Describe force as rate of change of momentum: $\text{force} = \dfrac{\text{change in momentum}}{\text{time taken}}$

$$F = \frac{(mv - mu)}{t}$$

Learning outcomes

- Be able to describe the operation of a number of safety features in vehicles in terms of extended stopping times
- Use the idea of rate of change of momentum to relate this to reduced forces

Resources

Student Book pages 56–60

Develop the activity: Video clips or photographs to illustrate the particular points being covered including crumple zones at the front and rear of a vehicle, seat belts and airbags.

Bring the activity together: Pictures of cars and pictures of children's playgrounds that have materials such as bark chippings or a specialised tile on the ground.

Approach

Introduce the activity

Ask students to work with a partner to summarise the main ideas from the previous activity. Explain that this lesson they are going to use exactly the same idea, but this time applied specifically to safety features in cars.

Develop the activity

Show some video clips or photographs to illustrate the particular points being covered – crumple zones at the front and rear of a vehicle, seat belts and airbags.

Ask students to produce display material to illustrate how each feature works – explaining their point each time with reference to change in momentum – i.e. the change in momentum is fixed so slowing the occupant down over a longer time reduces the force on them.

Provide some pictures of cars (e.g. from magazine advertising) so that the lesson is not held up by difficulties in drawing cars.

Ask students to do the worked examples on page 59, without the worked solution, then compare their solutions with the solutions in the Student Book.

Students can then tackle the questions on page 60 in pairs.

Bring the activity together

Show pictures of a children's playground that has a 'soft' material on the ground (could be a material such as bark chippings or a specialised tile). Ask students to explain how it reduces injuries to small children. Again the answer follows the same pattern – avoid answers that 'absorb the impact'.

A key feature across the lesson is that the same answer applies in a wide variety of situations – the students should be aware that exam questions can ask for this answer in a variety of ways.

Answers

Page 60

1. Momentum = mass × velocity. Momentum is a vector quantity (it includes a direction) because velocity is a vector quantity.

2. From Newton's second law, for a particular change in momentum, the longer the change takes the smaller the force will be. So if a parachutist bends their knees when they land, the landing takes a longer time, the momentum change is fixed and so the force is reduced.

3. 2.5 m/s.

4. 4 m/s.

5. a) 100 000 kg m/s.
   b) 100 000 kg m/s.
   c) 10 m/s.

6. When the ball hits the goalkeeper the padding changes shape and absorbs the kinetic energy of the ball. The ball will take a longer distance and time to stop. The change in momentum of the ball takes place over a longer time.

$$F = \frac{m(v - u)}{t}$$

If the time increases, then the force reduces, resulting in less injury.

Activity P1c.11 The third law

Specification reference: P1c.1.29P

Learning objectives

- Demonstrate an understanding of Newton's third law

Learning outcomes

- Describe the conditions for a pair of forces to be a third law pair
- Give simple examples of third law pairs

Common misconceptions

Students find it difficult to 'accept' ideas such as roads pushing cars forwards and tables attracting the Earth.

Resources

Student Book pages 61–63

Resources for class practical (see Technician's notes, following)

Approach

Introduction

Introduce the lesson by showing a video clip of a rocket taking off (or a picture if video is not available). Ask the students to explain why the rocket goes up. Answers may involve ideas about the engines providing a force which pushes the rocket.

Stronger answers may use ideas from the last two lessons and talk about the momentum change when the fuel is burned. Accept these answers and then point out that the engines produce a force downwards, so why does the rocket go up?

Newton's third law

Develop the lesson by introducing Newton's third law, as outlined on page 61. At this level explain that forces always come in pairs and state some of the conditions attached to those pairs – i.e. they are equal in size, opposite in direction and act on different objects.

Students find the application of this idea very difficult so give lots of simple examples, such as the rocket engines push downwards so the rocket is pushed upwards. However, be careful with your choice of examples – students find it difficult to 'accept' ideas such as roads pushing cars forwards and tables attracting the Earth, however true it might be, and this can lead to confusion.

Practical activities

Practical activities could involve demonstrating the effect using a balloon ('explain why the balloon flies around the room when I let it go') – the balloon can be attached to a guide string using a straw if you wish to investigate the acceleration produced.

Alternatively, commercial 'water rocket' kits are available which can be very engaging for students but potentially quite limited in their value as an experimental tool for gathering data.

Summary

Draw the lesson together by returning to the original video of a rocket launching and ask the students to describe how it works again – you are looking for accurate, sequenced answers at this point.

Technician's notes

Be sure to check the latest safety notes on these resources before proceeding.

The following resources are needed for the class practical P1c.11, per group:

balloon attached to guide string using a straw
water rocket kit

Answers

Page 61

1 Newton's third law states that the forces on each skater will be equal and opposite – so both experience a force of 200 N. Newton's second law states that $F = ma$. Rearranging the formula gives $a = F \div m$. So for the same force, the skater with more mass will have a lower acceleration. The accelerations are also in opposite directions.

Developing investigative skills, page 63

1. Make sure the ruler is perpendicular to the table top (by using a set square) and measure the height in several places across the ramp.

2. The student also needed to know the mass of the trolley as momentum = mass × velocity. The mass could be measured using a balance.

3. Independent variable: height of ramp; dependent variable: time take for trolley to go down ramp.

4. Graph.

5. As the height increases, the momentum increases. It is not directly proportional.

6. On their own, these data points would indicate a line, probably straight, giving a *y*-intercept of about 0.6 kg m/s (assuming momentum is on the *y*-axis).

7. You would expect the line to go through the origin as a ramp of starting height of zero will mean no momentum is gained by the trolley.

8. The student should make additional measurements between 0 cm and 15 cm to gather more data points and make the pattern in the line clearer in this region. (In practice, the student found it difficult to make these measurements as the friction tended to slow the trolley down too much).

Activity P1c.12 Consolidation and summary

Learning objectives

- To review the learning points of the topic
- To test understanding through answering questions

Learning outcomes

- Be familiar with the knowledge and understanding summarised in the End of topic checklist
- Be able to apply this knowledge and understanding by answering End of topic questions

Resources

Student Book pages 64–69

Approach

Introduce the activity

Ask students to work with a partner to make a list of key words from this topic. They could then work together to produce a spider diagram showing how the different concepts are linked. They could compare their list with the list of key words given on page 64 of the Student Book.

Discuss the checklist given on pages 64–65 and use questioning to see how much of the content students are comfortable with.

Develop the lesson

Ask students to work individually to work through the End of topic questions on pages 66–69 of the Student Book without reference to the text. As they work, walk around the classroom observing their answers and questioning them as necessary to find out which questions are causing difficulties.

Bring the lesson together

After a set period, ask the students to stop working and discuss any areas of difficulty you observed as you walked round the class.

Students should complete any unanswered questions for homework, but you should stress that they should attempt the questions without reference to the text to see how much content they have retained.

Answers

End of topic questions mark scheme

Question	Correct answer	Marks
1 a)	600 N	1
1 b)	60 kg	1
1 c)	228 N	1
1 d)	60 kg	1
1 e)	228 N	1
2 a)	stay similar because the force your muscles exert does not change	1 1
2 b)	increase because the force used to jump the same height is now less	1 1
2 c)	increase because the force used to vault to the same height is now less	1 1
2 d)	increase because the force used to throw the javelin the same distance is now less	1 1
2 e)	probably stay similar because speed depends on air resistance which has not changed [there is a case for decrease if, for example, jumping along turned out to be faster than running]	1 1
3 a)	stage 1 – the skydiver is accelerating. The downward weight is greater than the upward force caused by air resistance stage 2 – the skydiver is travelling at constant speed. The weight and air resistance are balanced stage 3 – the skydiver is slowing down. The force caused by the air in the parachute is greater than the weight stage 4 – the skydiver is travelling at a constant speed. The forces are balanced again	1 1 1 1

Question	Correct answer	Marks
3 b)	the weight is balanced by the upward force of the ground on the skydiver	1
4 a)	the graph is a straight line from the origin to the (12 cm, 6.0 N) point after that it become increasingly curved and less steep	1 1
4 b)	the limit of proportionality should be marked on the graph at the end of the region that is a straight line proportional behaviour occurs all along the region where the graph is a straight line. Plastic behaviour occurs thereafter, though not necessarily immediately	1 1
4 c)	if the spring returns to its original length then the behaviour was purely elastic. If it is longer, then the behaviour was partially elastic and partially plastic	1
5 a)	2.5 m/s^2	1
5 b)	2500 N	1
6	the fuel burns quickly and the hot gas expands and leaves the rocket at speed through the hole at the bottom this creates a downward force on the air and an equal and opposite upward force on the rocket the rocket accelerates and burns some the fuel which reduces the mass and increase the acceleration upward force is reduced when the fuel has burned which decreases the acceleration	1 1 1 1
7	800 Nm	1
8 a)	the stretching increases the time for the passenger to stop the force is reduced	1 1
8 b)	the seatbelt may be damaged and may not stretch again	1
8 c)	crumple zone collapsible steering wheel / column air bag	1 1 1

Question	Correct answer	Marks
9	Ahmed's clockwise moment is 300 N × 4 m = 1200 N m. Rod's anticlockwise moment is 450N × 4 m = 1800 N m. Freddy needs to provide a moment of 600 N m clockwise to balance the see-saw. Freddy weighs 300 N, so he needs to sit 2 m to the right of the pivot (300 N × 2 m = 600 N m)	1 1
10	acceleration = rate of change of velocity as changing direction changes the velocity	1
11	the car decelerates / slows down when it hits the wall the dummy continues forward at 35 km per h the dummy hits the dashboard / windscreen / steering wheel and experiences a large force	1 1 1
12	balanced = steady speed no resultant force	1 1
13	unbalanced – slowing down so there is a resultant force	1 1
14	acceleration = 2 m/s^2, resultant force = 3000 N this is an average because the actual force will have varied over time as the frictional forces varied	1 1
15	resultant force = force from engine – frictional forces (opposing motion) 3000 N = force from engine – 200) N force from engine = 5000 N	1 1
16	lower gravitational field strength so time of flight is longer as vertical acceleration is lower also lack of air resistance means no horizontal deceleration	1 1 1
17	gravity causes drops to accelerate down. as the speed increases, drag increases until resultant force is zero at this point raindrop falls at constant speed (terminal velocity) so raindrop gains no further kinetic energy	1 1 1

Question	Correct answer	Marks
18	list of factors on pages 42–43 with explanations as to why they make a difference in changing the reaction time (factors to do with the driver) and those relating to friction (brakes, road surface etc)	3 marks max (1 mark per correct factor and explanation)
19 a)	7500 kg m/s	1
19 b)	1500 N	1
19 c)	forces of friction between the car and the road, air resistance	1
20	the skateboard can move so there must be the same momentum going backwards as there is going forwards	1 1
21	moment = force × distance from pivot, so pushing opposite to the hinge increases the distance and less force is needed to get the same moment	1 1
22 a)	100 m, average speed = $d \div t$ = 100 ÷ 9.58 = 10.44 m/s 200 m, average speed = $d \div t$ = 200 ÷ 19.19 = 10.42 m/s 400 m, average speed = $d \div t$ = 400 ÷ 43.03 = 9.30 m/s 800 m, average speed = $d \div t$ = 800 ÷ 100.91 = 7.93 m/s 1500 m, average speed = $d \div t$ = 1000 ÷ 131.96 = 7.58 m/s	1 1 1 1 1
22 b)	the runners spend time accelerating for 100 m then a short distance cruising at top speed / AW At 400 m (for example) they spend a similar time accelerating but then cruise for a longer distance but at a slower speed / AW	1 1
23	yes, if the graph of force against extension is a straight line rubber will follow the proportional path until the polymer chains are more straightened out however a sample of rubber may need a little extra force to get started and so not follow the straight line all the way to the origin	1 1 1

Question	Correct answer		Marks
24	stiff springs: reduce body roll, go round corners faster soft springs: absorb the bumps better, more comfy ride		1
	so choose stiff springs		1
25	This is a level of response question. For example:		
	Level 1:	Car increases its speed (or accelerates) steadily.	2
		Or car has constant acceleration.	2
		Or car increases speed.	1
	Level 2:	Car accelerates steadily at 0.5 m/s^2 for 4 s, for 0 to 2.0 m/s, or moves a distance of 4 m.	4
		Or car has a constant acceleration of 0.5 m/s^2.	3
	Level 3:	Car accelerates steadily at 0.5 m/s^2 for 4 s, from 0 to 2.0 m/s, and moves a distance of 4 m.	6
		Or car has a constant acceleration of 0.5 m/s^2 and moves a distance of 4 m in 4 s.	5
	Total:		67

Section 1: Answers to exam-style questions

Question	Correct answer	Marks
2 a) i)	speed = distance ÷ time (or equivalent)	1
2 a) ii)	speed = 100 ÷ 9.58 10.4 (m/s) 3 s.f. only	1 1 1
2 a) iii)	this is an average he would have run slower at the start so he would have run faster than the average at some point	1 1 1
2 b)	the runner must react to the start it takes time (0.1 s minimum) for brain to process/get muscles to move	1 1 1
3 a) i)	between 10s and 30s	1
3 a) ii)	because the graph is horizontal	1
3 b)	acceleration = gradient 8 ÷ 10 = 0.8	1 1 1
3 c)	area under graph = distance; (1/2 × 10 × 8) + (20 × 8) + (1/2 × 15 × 8) 260 (m)	1 1 1
4 a)	tape measure stopwatch	1 1
4 b)	do not step onto road / stay a safe distance from the roadside	1

Question	Correct answer	Marks
4 c)	measure distance in metres fixed distance / distance between two fixed points measure time with a stopwatch / stop clock (measure) time taken for cars to travel the fixed distance	1 1 1 1
4 d)	column headings are: Cars Distance moved by cars (m) Time taken (s)	 1 1 1
4 e)	calculate the speed for each vehicle by using speed = distance ÷ time compare the speeds to see how many cars exceed 14 m/s calculate the % that are breaking the speed limit more cars in the sample will increase confidence in the conclusion	1 1 1 1
5 a)	initially momentum is zero Alex has momentum in one direction after so Tom must have a negative momentum so that final momentum is zero overall	1 1 1 1
5 b) i)	momentum = mass × velocity (or equivalent)	1
5 b) ii)	50 × 1.5 75 kg m/s	1 1 1
5 b) iii)	– 75 (kg m/s)	1 1
5 b) iv)	75 ÷ 1.2 62.5 kg	1 1

Question	Correct answer	Marks
5 c)	the force must be equal magnitude opposite in direction	1 1
6 a) i)	$W = m \times g$	1
6 a) ii)	2000 × 6 12 000 (N)	1 1
6 b) i)	container gives clockwise moment concrete block provides anticlockwise moment to provide balance	1 1 1
6 b) ii)	clockwise moment = anticlockwise moment (12 000 × 12) = (W × 3) W = 48 000 (N) mass = 4800 kg	1 1 1 1
7 a)	hold vertically zero next to start of spring head at same height to avoid parallax errors	1 1 1
7 b)	any ONE suitable, e.g. safety glasses in case spring flies up aware of risk of masses falling	1
7 c) i)	suitable scales plotting points suitable line	2 2 1
7 c) ii)	anomaly identified does not fit pattern	1 1
7 c) iii)	yes straight line through origin	1 1 1
	Total	70

Section 2. Electricity

Contents

Overview of the section

This section begins by considering mains electricity. Students are reminded about the hazards associated with it, and the use of double insulation, earthing, fuses and circuit breakers. The first topic introduces the relationship between power, current and voltage, and energy transferred, current, voltage and time. The first topic ends with a consideration of the differences between alternating current (a.c.) and direct current (d.c.).

The second topic considers energy and voltage in circuits. It begins by comparing series and parallel circuits. Ohm's law is used in various contexts. Students are reminded that current is a flow of charge. The relationship between charge, current and time is introduced and used. Finally, students learn that voltage is the energy transferred per unit charge.

The section ends with a topic that considers electric charge. It begins by considering the difference between electrical conductors and insulators, before describing how materials can become charged by friction. Electrostatic phenomena are explained in terms of electrons and the hazards and uses of electrostatics are explored.

Starting points

The Student Book provides a 'section opener', a double-page spread which sets the scene for students (see pages 76–77). It is structured in this way:

- an overview providing details of the areas of study
- six questions, questions 1, 2, 3 and 6 relate to aspects of physics that should be reasonably familiar to most students, and questions 4 and 5 relate directly to some of the topics to be studied
- a list of section contents.

The questions provide a structure for introducing the section.

1. Two lamps, A and B, are connected in series with a battery. Lamp A is unscrewed from its holder, which causes a break in the circuit. What will happen to lamp B? Why?

2. Two lamps, A and B, are connected in parallel with a battery. Lamp A is unscrewed from its holder. What will happen to lamp B? Why?

3. What happens to electric current as it goes round a circuit?

4. A series circuit consists of two ammeters, A1 and A2, and two bulbs connected to a battery. Which ammeter will have the biggest reading? Why?

5. Name some safety devices used with mains electricity. Can you explain how each one works?

6. Suggest why different electrical appliances in the home use different voltages and make a list of some common examples and their voltages.

The section opener has two main purposes:

- to acknowledge the student's prior learning and to value it
- to provide a benchmark against which future learning can be compared.

The six questions or 'starting points' in the Student Book can be used in a number of different ways to introduce the section:

- You could ask students to consider the questions as an introductory homework task.
- You could put students into groups to share their own ideas and understanding and then report back to the whole class.
- Students could be given access to the Internet, preferably with a tight timescale, to find out the information required.

You could then use a spider chart or other form of wall chart to summarise everybody's ideas.

Recording these initial ideas in this way allows you to retain them for reference as the individual topics are developed. In this way, your students' progress in learning can be readily acknowledged.

P2b Mains electricity

Introduction

In this topic students learn about mains electricity. They learn about how the hazards associated with mains electricity are mitigated by safety features. They learn about electrical heating and power, and fuse ratings. Finally, a.c. and d.c. supplies are considered.

Links to other topics

Section	Essential background knowledge	Useful links
2 Electricity		P2c Energy and voltage in circuits P2d Electric charge
4 Energy resources and energy transfers		P4d Energy
6 Magnetism and electromagnetism		P6d Electromagnetic induction

Topic overview

P2b.1	**Domestic electrical safety features** Students learn about electrical safety features in a domestic context.
P2b.2	**Electrical heating and current in a resistor, power, voltage, current and time** This activity defines electrical power and its unit. It explains the relationship between power, current and voltage. It describes examples of electrical energy being used to supply a heating effect. It explores the relationship between energy, current, voltage and time.
P2b.3	**a.c. and d.c.** This activity describes the difference between an a.c. supply and a d.c. supply and gives a simple explanation of why a.c. is used for mains supply.
P2b.4	**Consolidation and summary** This activity provides an opportunity for a quick recap on the ideas encountered in the topic as well as time for the students to answer the End of topic questions in the Student Book.

Activity P2b.1 Domestic electrical safety features

Specification reference: P2b.2.2

Learning objectives

- Explain the use of a number of safety features (insulation, double insulation, earthing, fuses and circuit breakers) to prevent harm to appliances and users

Learning outcomes

- Explain why safety is an important consideration in the use of electrical appliances
- Describe the operation of safety features such as double insulation, fuses, circuit breakers and earth wires
- Describe the sequence of how a fuse acts to make a circuit safe

Common misconceptions

Often students think of 'electricity' as a 'thing' that moves around in wires – discourage the use of the word 'electricity' in everything except the topic title. This will help to remove the confusion as the electron model is built up during the topic.

Resources

Student Book pages 79–82

Worksheet P2b.1a Fuse blowing

Resources for class practical and demonstration (see Technician's notes, following)

Approach

Introduce the activity

You could ask the class to tell you what they know about electrical safety features in the home, generating a concept map or spider diagram on the whiteboard. Don't correct any errors – leave that to the plenary.

Introduce the class to the residual current circuit breaker (RCCB) in the laboratory. Demonstrate its use, explain its presence and discuss how it works.

Develop the activity

Gather information about our everyday use of electricity. This could be through discussion in small groups, describing electrical appliances from home, or through a survey of the room or building.

Discuss insulation, fuses and circuit breakers. Discuss the text on pages 79–80 of the Student Book. Students then carry out the practical on Worksheet P2b.1a. Check that students know what the different components in the circuit are for and how they are connected.

Students connect the fuse, meter and lamp in series. They slowly increase the voltage from the power supply and observe what happens to the lamp, meter and fuse. Take care as the fuse is very hot. The lamp needs to be rated at 12 W or 24 W, so that the fuse blows before the lamp does.

Subdued lighting makes it easier for the students to observe the glow of the fuse just before it melts. Students often rush the practical and fail to observe key details. You might want to have two fuses for each group.

Earth safety demonstration

Students could observe the earth safety demonstration described below. Introduce the RCCB and explain its advantages over simple fuses and circuit breakers.

Assemble the circuit, explaining the role of each cable as it is connected.

The loose wire can be left out at this stage.

Coil the fuse wire on the heatproof mat so that it isn't illuminated by the lamp of the ray box.

The earthing of the ray box casing could be done through a second power pack.

Turn on the a.c. supply so that the lamp shines brightly.

The fuse should get slightly warm.

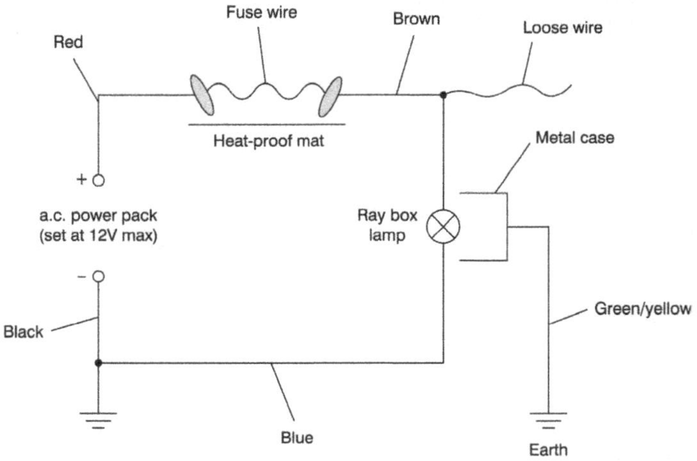

Turn the power off to add the lead. Make sure you only handle the insulated part of the lead when touching it to the metal casing.

Now add a red lead to the circuit somewhere between the fuse and the lamp. Explain why this is a potentially lethal situation.

Touch the other end of the red lead to the casing and watch the fuse blow and cut the lamp off from the supply. Take care as the fuse will be very hot.

You might want to repeat with the casing not earthed.

All students should tackle the questions on page 82 in pairs.

Bring the activity together

Use class questioning to correct/improve the concept map or spider diagram generated in the starter. Issue whiteboards and marker pens. Set them some general questions about connecting an appliance to the mains, then finish off with a fuse rating calculation or two.

Technician's notes

Be sure to check the latest safety notes on these resources before proceeding.

The following resources are needed for the class demonstrations, P2b.1:

Introduction

access to the laboratory's RCCB

Earth safety

12 V a.c. supply with earthing point
ray box lamp with metal outer casing
coil of 5 A fuse wire
two red 4 mm leads, black 4 mm lead, green or yellow 4 mm lead, brown 4 mm lead, blue 4 mm lead
heatproof mat
two crocodile clips

The power pack needs to supply at least 6 A to ensure that the fuse blows.

Check that the voltage used is sufficient to blow the fuse before the activity.

The following resources are needed for the class practical P2b.1a, per group

20 mm glass bodied fuse rated at 800 mA
12 V ray box lamp or similar
ammeter (0–1 A)
four 4 mm leads
two crocodile clips
variable d.c. power supply for ray box

The lamp needs to be rated at 12 W or 24 W, so that the fuse blows before the lamp does. Subdued lighting makes it easier for the students to observe the glow of the fuse just before it melts.

Take care as the fuse is very hot.

Answers

Page 82

1. Brown wire: live, connected to fuse. Blue wire: neutral wire. Green and yellow wire: earth wire.

2. It melts when the current gets too high.

3. The student should not choose the 'nearest' fuse, but the 'nearest above'. If the appliance requires 6 A, the 5 A fuse will break whenever the appliance is used.

4. The earth connection needs to be a low resistance path (see later in the chapter). This means that a high current will pass through the wire and this will melt the fuse (or trip the circuit breaker).

5. The casing cannot become live because it is not a conductor.

P2b.1a Fuse blowing

You are going to set up a circuit and watch a fuse blowing.

Apparatus

power pack

ray box lamp

leads

ammeter

fuse

crocodile clips

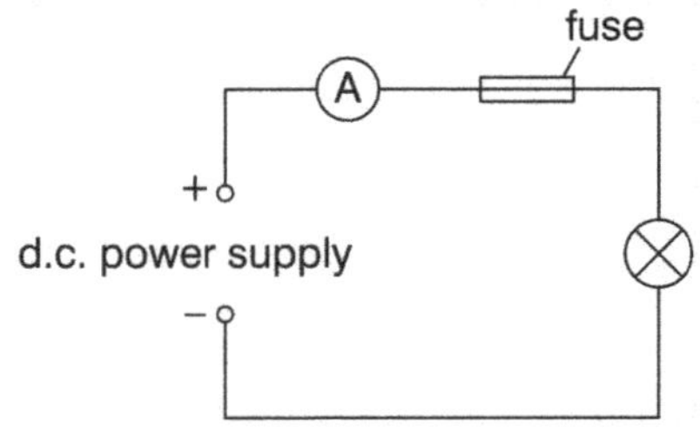

SAFETY INFORMATION

Take care as the fuse is very hot.
The ray box may get hot.

Method

1. Turn the power pack down to 0 V.
2. Assemble the circuit as shown above.
3. Increase the voltage on the power pack **very** slowly, observing the ammeter, lamp and fuse as you do.

Activity P2b.2 Electrical heating and current in a resistor, power, voltage, current and time

Specification reference: P2b.2.3, P2b.2.4, P2b.2.5

Learning objectives

- Understand why a current in a resistor results in the transfer of electrical energy and an increase in temperature
- Describe examples of electrical energy being used to supply a heating effect in a domestic context
- Define electrical power and its unit: power = current × voltage
  $P = I \times V$
- Use the relationship between energy transferred, current, voltage and time: energy transferred = current × voltage × time
  $E = I \times V \times t$

Learning outcomes

- Describe examples of electrical heating
- Recall that power is the energy transferred per second and is measured in watts
- Be able to recall and use the equation $P = I \times V$
- Calculate the current required for an appliance given its power rating and operating voltage
- Choose the correct value of fuse for an appliance
- Use the equation $E = I \times V \times t$

Common misconceptions

Students will not yet be used to dealing with mixed units in science, so will need to be made aware of the need to get values expressed in basic units (W, J or s) before putting them in a formula.

Students often believe that the fuse protects the appliance. Whilst this is indirectly true, it is often more likely that any danger of overheating or even catching fire is going to be in the connecting cables, rather than in the device itself. Many students gain the idea that the fuse somehow actively stops the current getting too high and so keeps the current at the correct level. This is not true – if the current gets too high the fuse melts and breaks the circuit. 'It stops the current getting too high' is not a good phrase to use as many students misunderstand this as the fuse actively keeping the current low.

Resources

Student Book pages 82–87

Worksheet P2b.2a Kettle power

Resources for class demonstrations (see Technician's notes, following)

Approach

Introduce the activity

Arrange a circus of mains electrical appliances around the room. Give students ten minutes to find out the power rating (W or kW) of each appliance from its base plate. Explain the significance of the power rating with an energy transfer calculation, involving a power rating in kW and a time in minutes or hours to emphasise the need to get quantities into base units before doing calculations.

It is safer to turn off the mains supply to the room while students are handling the appliances.

Present students with a microwave oven. Invite one of them to find the base plate and tell you what its power rating is. Place a potato in the oven and use a stop watch to measure how long it takes to cook it. Show them how to combine both numbers to calculate the electrical energy transfer down the mains lead.

Develop the activity

Discuss 'Electrical heating and current in a resistor' on page 82 of the Student Book. Students should answer the questions on page 83 of the Student Book in pairs.

Now move on to discuss the link between power, current and voltage. Remind students what they observed in the introduction to the activity.

Allow students to do the worked examples on page 84 of the Student Book without the given solution. Then compare students' solutions with the given solutions and discuss any differences. Students should attempt the questions on pages 85–86 (top) of the Student Book with a partner.

Now carry out the kettle demonstration. You use a kettle to boil a known mass of water. Boiling water is scalding, and appropriate care should be taken.

Students should be encouraged to help with measurement taking, but no measurements of the temperature of the boiling water should be made as this is too hazardous. The measurements you need to find the heating power of the kettle are:

- mass of empty kettle
- mass of kettle + water
- mass of water
- temperature rise of water
- time taken.

From this students calculate the power rating of the kettle. Students can use Worksheet P2b.2a to carry out the calculation.

Students should attempt the worked example on page 86 of the Student Book as described above before they continue. Students will not only need to be able to transpose the formula $E = P \times t$ to do these questions but also deal with conversions of power and time to basic units. Some of them will find this difficult, but you could point out that this is what people have to do in the real world of work – not just scientists.

Students can then do the questions on page 86 (bottom) of the Student Book in pairs.

Bring the activity together

Ask students to write down two things that they did well in the activity and one thing that they feel they could improve. Discuss responses and deal with any issues that arise.

Technician's notes

Be sure to check the latest safety notes on these resources before proceeding.

The following resources are needed for the class demonstrations, P2b.2.

Introduction

A range of mains electrical appliances with base plates that shows their power rating, which could include:

- kettle
- TV
- kitchen blender
- hairdryer
- desk lamp
- computer
- microwave oven and potato
- stop clock

It is safer to turn off the mains supply to the room while students are handling the appliances.

Place a potato in the oven and use a stop watch to measure how long it takes to cook it.

Kettle power

plastic kettle (white if possible)
stop clock
scales
thermometer

Boiling water is scalding, and appropriate care should be taken.

Students should be encouraged to help with measurement taking, but no measurements of the temperature of the boiling water should be made as this is too hazardous.

Answers

Page 83

1. A resistor is a device that opposes the flow of current.
2. It increases.
3. Electrical energy to heat energy.

Pages 85–86 (top)

1. 220 W.
2. 12 V.
3. 0.11 A.
4. a) 1.6 A, 3 A fuse.
   b) 3.47 A, 5 A fuse.
   c) 0.4 A, 3 A fuse.
   d) 9.1 A, 13 A fuse.
   e) 3.9 A, 5 A fuse.

Pages 86–87

1. The charger will still work, although the battery may not charge as effectively. The transformer in the charger (see Section 6) will reduce the mains voltage by the same factor, so the battery may not be able to charge fully.
2. The charger will reduce the mains voltage by the same factor as usual, but starting with 230 V, the resulting voltage may be high enough to damage the circuitry in the laptop.
3. 2160 J.
4. 3.3 A.
5. 218 s.
6. 12 V.

P2b.2a Kettle power

1. Complete these calculations to find the heating power of the kettle.

Mass of empty kettle =_____kg

Mass of kettle + water = _____ kg

Mass of water = _____ kg

Initial temperature of water = _____ °C

Temperature rise of water = 100 °C – _____ °C = _____ °C

Energy transferred (J) = mass (kg) × temperature rise (°C) × 4200 J/kg °C

Energy transferred = _____ kg × _____ °C × 4200 J/kg °C = _____ J

Time taken = _____ s

$$\text{Power rating} = \frac{\text{energy transferred (J)}}{\text{time taken (s)}} = \frac{\_\_\_\ \text{J}}{\_\_\_\ \text{s}} = \_\_\_\_\ \text{W}$$

Kettle power rating = _____ kW

2. Why do you think the measurement of the power rating does not agree with the rating on the kettle? Explain your answer below.

Activity P2b.3 a.c. and d.c.

Specification reference: P2b.2.6

Learning objectives

- Know the difference between an a.c. supply (mains electricity) and a d.c. supply (a cell or battery)
- Explain why a.c. is used for mains supply

Learning outcomes

- Describe the difference between a.c. and d.c. in terms of electron movement
- Draw sketch graphs to illustrate the difference over time of the two methods

Resources

Student Book pages 87–88

Worksheet P2b.3a a.c. to d.c.

Resources for class demonstration and practical (see Technician's notes, following)

Approach

Introduce the activity

Use the following alternating current demonstration to start the activity. Assemble this circuit to start with.

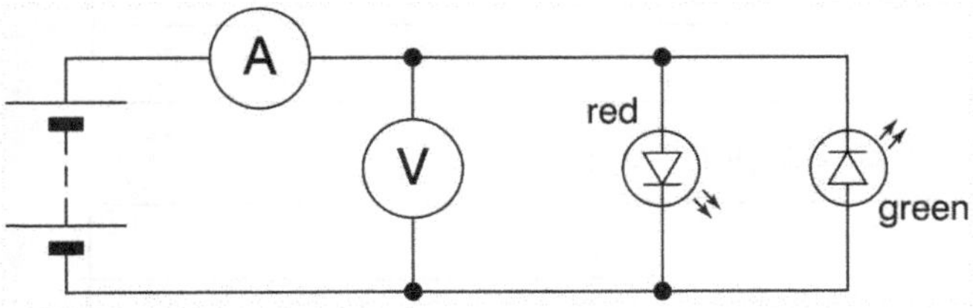

Use d.c. to explain why only the red LED glows. Swap round the terminals of the battery so that only the green LED glows.

Then replace the battery with the signal generator set to 0.1 Hz (or slower) and 4 V RMS. Relate the behaviour of the signal generator to the behaviour of the meters and LEDs.

Increase the frequency until both LEDs appear to glow continually and the meters read zero.

Replace the voltmeter with an oscilloscope to display a voltage–time graph of the voltage across the LEDs. Vary the frequency and voltage of the signal generator to show what effect it has on the oscilloscope display.

Although a cathode ray oscilloscope works fine, an analogue-to-digital converter connected to a laptop computer and a digital display can be more effective. Converters (such as PicoScope) are easily available from educational suppliers.

Develop the activity

Discuss the text on page 87 of the Student Book which describes the difference between alternating current (a.c.) and direct current (d.c.).

At this point the students can carry out the practical on Worksheet P2b.3a which shows them how alternating current from a mains power supply can be converted into steady direct current for components which need it. Students connect the motor to the a.c. supply with the ammeter in series and the voltmeter in parallel.

They note the effect of adding the diode in series with the supply, and the effect of reversing the diode. They then replace the motor with the LED and note the smoothing effect of the capacitor.

Students can study the Science in context feature on pages 87–88 and make their own summary of it in a form of their choice.

Bring the activity together

Ask students to make their own summary of the main points of the activity.

Technician's notes

Be sure to check the latest safety notes on these resources before proceeding.

The following resources are needed for the class demonstration P2b.3:

signal generator with high current output
two LEDs with built-in 100 Ω series resistors, one red, the other green
leads
centre-zero moving coil ammeter (0–50 mA)
centre-zero moving coil voltmeter (0–5 V)
display oscilloscope
6 V battery

The following resources are needed for the class practical P2b.3a, per group:

3 V a.c. power supply
3 V d.c. motor
diode
voltmeter (0–5 V)
ammeter (0–1 A)
capacitor (at least 1000 μF)
LED with built-in protective resistor
leads
eye protection

A silicon diode rated at 3 A or more will be fine.

A capacitor of value 1000 μF which is rated at 16 V should be safe – students need to know which way round to connect it.

You could provide pairs of capacitors permanently connected in series with each other by their positive terminals – then it doesn't matter which way round they go.

The eye protection is for the unlikely event that the capacitor explodes when it is connected round the wrong way.

A 100 Ω resistor in series with the LED should be fine.

P2b.3a a.c. to d.c

This practical shows you how a.c. from a mains power supply can be converted into steady d.c. for components which need it.

Apparatus

a.c. power supply

d.c. motor

diode

capacitor

voltmeter

leads

ammeter

LED

eye protection

Method

1. Wearing eye protection, set up this circuit. Verify that nothing much appears to happen when the supply is turned on.

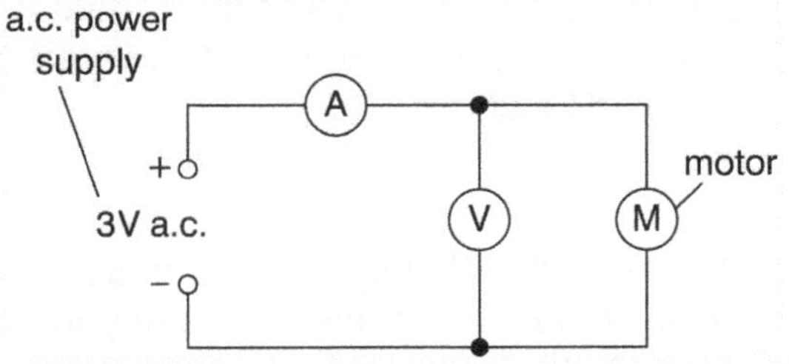

2. Now add the diode as shown below. Note what effect it has.

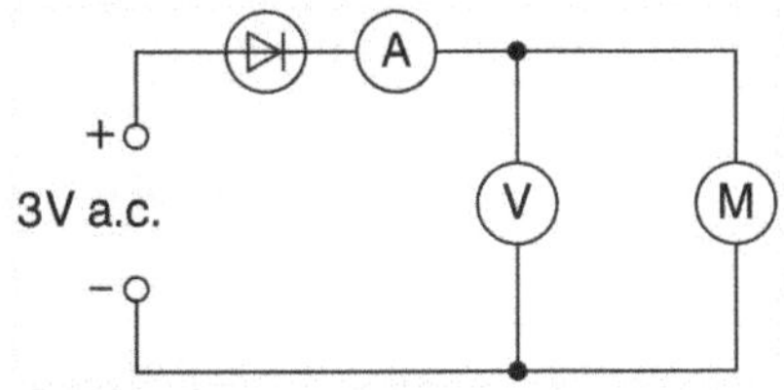

3. Find out what happens when the diode is turned round as shown below.

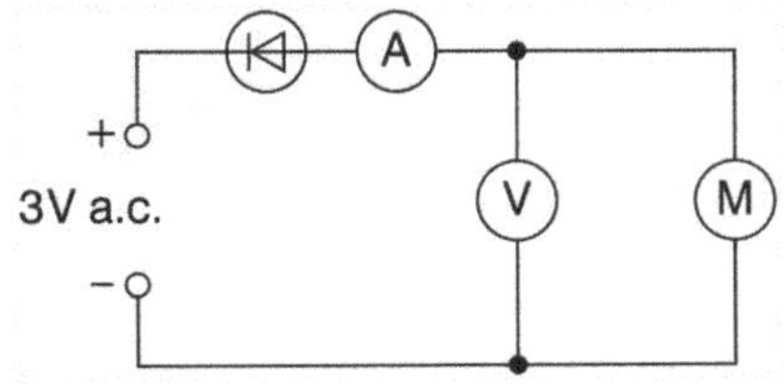

4. Turn the diode round again. Replace the motor with an LED as shown below. You should be able to see that it is flickering. Can you explain why?

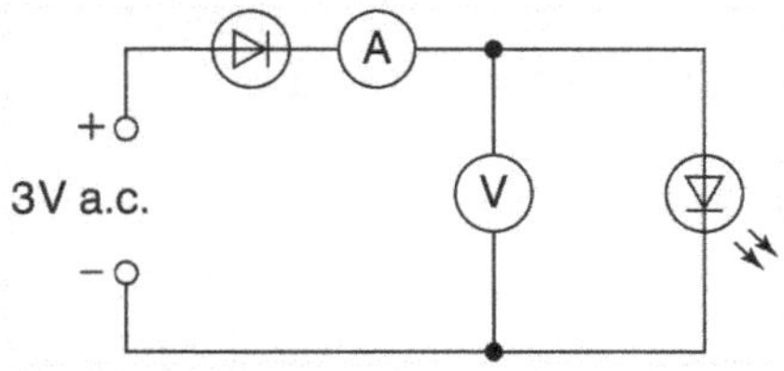

5. Now add a capacitor (at least 1000 μF) to smooth the voltage across the LED. Make sure the capacitor is connected round the right way. The LED should no longer flicker, and the readings on the meters should increase.

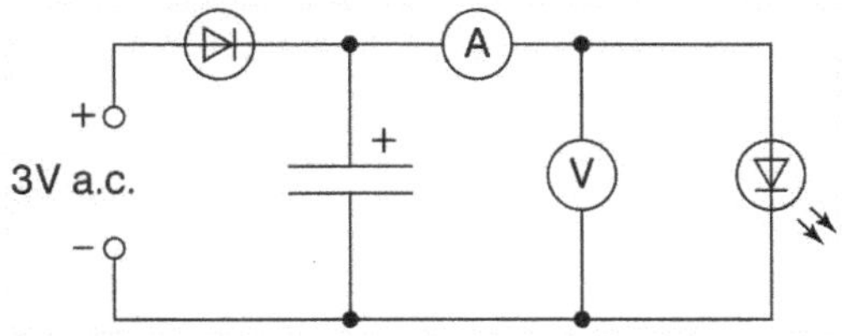

Activity P2b.4 Consolidation and summary

Learning objectives

- To review the learning points of the topic
- To test understanding through answering questions

Learning outcomes

- Be familiar with the knowledge and understanding summarised in the End of topic checklist
- Be able to apply this knowledge and understanding by answering End of topic questions

Resources

Student Book pages 89–91

Approach

Introduce the activity

Ask students to work with a partner to make a list of key words from this topic. They could then work together to produce a spider diagram showing how the different concepts are linked. They could compare their list with the list of key words given on page 89 of the Student Book. Discuss the checklist given on pages 89–90 and use questioning to see how much of the content students are comfortable with.

Develop the lesson

Ask students to work individually to work through the End of topic questions on page 91 of the Student Book without reference to the text. As they work, walk around the classroom observing their answers and questioning them as necessary to find out which questions are causing difficulties.

Bring the lesson together

After a set period, ask the students to stop working and discuss any areas of difficulty you observed as you walked round the class. Students should complete any unanswered questions for homework, but you should stress that they should attempt the questions without reference to the text to see how much content they have retained.

Answers

End of topic questions mark scheme

Question	Correct answer	Marks
1	if the wire is carrying a larger current, the diameter must be bigger so that the wire does not get too hot	1 1
2	only the live has a big enough voltage to kill. Once the fuse has blown, the supply voltage is disconnected, making the appliance safe to touch	1
3 a)	920 W	1 1
3 b)	water greatly increases the hazard of any contact with mains electricity as the skin becomes highly conducting if the hairdryer is dropped in the washbasin, then any attempt to pick it out will be fatal	1 1
4	current is 6.1 A so a 13 A fuse should be used	1 1 1
5 a)	you should fit an MCB of 30 A	1 1
5 b)	you should fit an MCB of 40 A	1 1
6	900 J	1 1
7 a)	5 A	1 1
7 b)	13 A	1 1
8	current that flows is 6.67 A so the fuse is rather high in value	1 1
9	0.05 A	1 1
10	352 W	1 1
11	111 V	1 1
	Total:	28

P2c Energy and voltage in circuits

Introduction

In this topic, students look in detail at electric circuits. They consider series and parallel circuits before looking at the relationship between voltage, current and resistance. They learn that current is a rate of flow of charge.

Links to other topics

Section	**Essential background knowledge**	**Useful links**
2 Electricity	P2b Mains electricity	P2d Electric charge
6 Magnetism and electromagnetism		P6c Electromagnetism P6d Electromagnetic induction

Topic overview

P2c.1	**Series and parallel circuits** In this activity students recall the symbols used to draw circuit diagrams and describe what happens to the current and voltage in series and parallel circuits.
P2c.2	**How changing resistance affects current** This activity explains the circuit operation for an ohmic conductor (e.g. resistance wire). It also considers the behaviour of non-ohmic devices such as thermistors and LDRs.
P2c.3	**Voltage, current and resistance** This activity explores the relationship between voltage, current and resistance.
P2c.4	**Voltage, energy and charge** This activity explores the relationship between voltage, energy and charge.
P2c.5	**Consolidation and summary** This activity provides an opportunity for a quick recap on the ideas encountered in the topic as well as time for the students to answer the End of topic questions in the Student Book.

Activity P2c.1 Series and parallel circuits

Specification reference: P2c.2.7

Learning objectives

- Describe what happens to the current and voltage in series and parallel circuits
- Explain why series or parallel circuits are appropriate in particular circumstances

Learning outcomes

- Recall the symbols used to draw circuit diagrams
- Describe the relative merits of series and parallel circuits in simple situations (e.g. lighting circuits)

Common misconceptions

Many students find assembling parallel circuits from a circuit diagram difficult. The trick is to assemble the series parts first and then put in the parallel ones.

Students want something to be 'used up' as it goes around a circuit. This gets in the way of their accepting the idea of conservation of current in a series circuit. It helps to emphasise from the start that the current carries energy which is transferred on its way round the circuit.

Students need to have the correct model of current as electron flow to appreciate why current at a junction is conserved. The idea that components in parallel have the same voltage is easy to grasp once students understand that voltage is what a battery confers on electrons.

Resources

Student Book pages 92–97

Worksheet P2c.1a Series circuits

Worksheet P2c.1b Parallel circuits

Resources for demonstrations and class practicals (see Technician's notes, following)

Approach

Introduce the activity

Display the circuit shown below on the board. Assemble the circuit methodically, talking about the function of each component as you put it in place. Point out the polarity of the LED. Leave out the voltmeter to start with. Explain that the blobs in the diagram show where leads must be stacked to obtain the parallel connection.

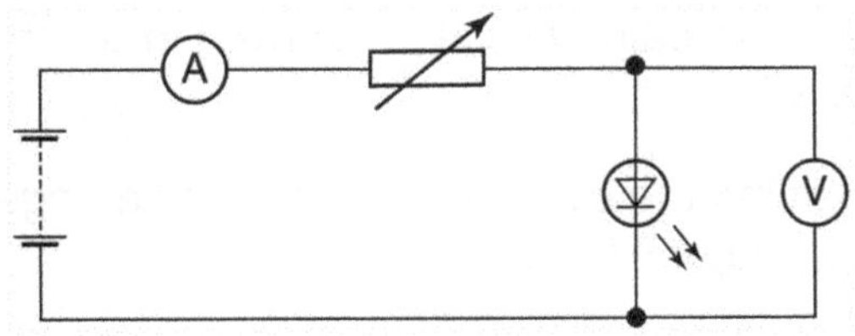

Display the following circuit on the board. Say what it does (the LED glows if enough light hits the LDR). Assemble the circuit using the method described above. Demonstrate the circuit in action. Ask for suggestions of applications for it.

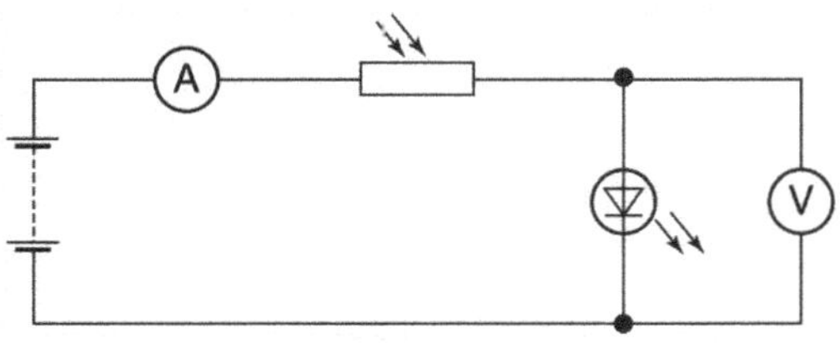

Develop the activity

Move on to discuss series and parallel circuits. Get the whole class to role play a simple circuit. They need to form a loop, perhaps around some benches, so that each student represents a free electron. Set up a tray of objects (wrapped sweets, coins) at one point of the loop. Label it 'battery'.

Have another tray labelled 'lamp' diametrically opposite. Students march round clockwise, picking up a piece of energy from the battery and depositing it in the lamp before returning to the battery. March them faster to increase the current.

Let them pick up two pieces of energy to increase the voltage of the battery. Point out that current is the same all the way round. It is worthwhile emphasising at this stage that the voltage across a component is a measure of how much energy is transferred by the current, so it can get transferred from the electrons to the components on its way round the circuit.

Students then carry out the practical on Worksheet P2c.1a.

Now connect a single lamp to a mains power pack so that it glows with normal brightness. Then connect a second lamp in parallel. Use class questioning to establish that because both lamps have the same brightness as the original one, they must have the same current and voltage.

Then establish that each electron which leaves the negative terminal of the power supply must pass through one of the lamps but not through both. Challenge the class to decide how the current in the power supply is related to the current in each lamp.

At this stage the students can carry out the practical on Worksheet P2c.1b.

Students connect the resistors in parallel with the battery and then place the ammeter in different places to establish that the current in and out of the battery is the sum of the currents in each resistor.

Less able students often instinctively build parallel circuits as though they were series ones, so it is advisable to keep an eye on the circuits that they assemble. A useful strategy is to get less able students to build an initial series circuit for one component, then add on the parallel parts one at a time so they can see them added on 'across' the original circuit.

Discuss the differences between parallel and series circuits as described in the table on page 94 of the Student Book. Students should tackle the questions which follow in pairs. Discuss 'Current in a series or parallel circuit' and 'How much current in the circuit?' on page 95.

Demonstrate how to measure how current varies with voltage using the circuit in Figure 2.23 on page 96 of the Student Book. Students should then tackle the questions on page 97 in pairs.

Bring the activity together

Provide the class with one power supply, two lamps and three switches. Challenge them to guide you through assembling a circuit where each lamp can be turned on and off independently, but only when the master switch is closed. Here is the circuit diagram:

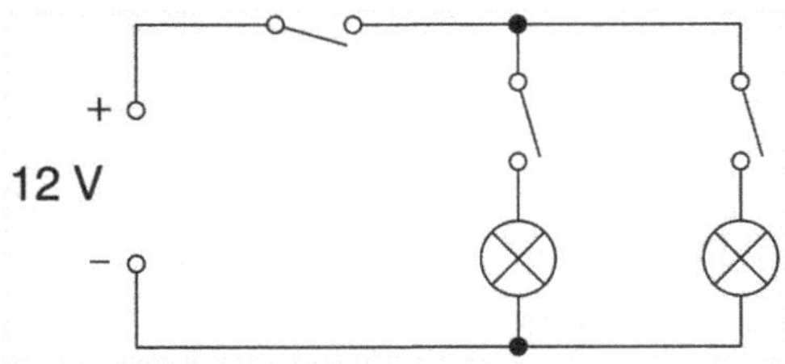

Technician's notes

Be sure to check the latest safety notes on these resources before proceeding.

The following resources are needed for the class demonstrations P2c.1.

Introduction

6 V battery
ammeter (0–20 mA)
voltmeter (0–2 V)
variable resistor (0–1 kΩ)
LED (with inbuilt 100 Ω safety resistor)
LDR
torch
eight 4 mm leads

The LED is best seen in subdued lighting.

Main part of the activity

two trays, one labelled battery and the other labelled lamp
coins or wrapped sweets equal to exactly twice the number of students
two 12 V ray box lamps
mains 12 V d.c. power supply
four 4 mm leads

End of the activity

12 V power supply
two 12 V ray box lamps
three press-to-make switches
nine 4 mm leads

The following resources are needed for the class practical P2c.1a, per group:

battery
three lamps
ammeter (0–1 A)
voltmeter (0–5 V)
seven 4 mm leads

A 2 W power rating for the resistors should be enough to stop them getting too hot.

Warn students not to touch them until they have had a chance to cool down a little.

The following resources are needed for the class practical P2c.1b, per group:

3 V battery
ammeter (0–1 A)
22 Ω resistor and 10 Ω resistor
seven 4 mm leads

Take care! If you use thick wire, the current will be high for a low voltage. Wire can get very hot very quickly and there is a risk of injury.

Answers

Page 95

1. There is probably a break in the circuit.

2. In a parallel circuit, more energy is transferred each second – the bulbs are brighter. The energy available from the battery is limited by the chemicals used – running two bulbs at once at full power will transfer the chemical energy more quickly and the battery will become flat sooner.

3. So that they can be switched independently.

4. a) In a series, dimmer because battery energy shared between two bulbs.

   b) In parallel, full energy of battery given to each bulb.

Page 97

1. Bulb B has the higher resistance. The brighter bulb must have the higher current (since, in parallel, both bulbs are connected to the same voltage), so it must also have the lower resistance.

2. In series, bulb B will be brighter. It has a higher resistance, so in series it will have a bigger share of the voltage and hence the energy.

3. Use a circuit like the one shown in Fig. 2.22. Vary the voltage and record the current flowing in the circuit.

4. The ammeter should be connected in series and voltmeter in parallel with the component.

P2c.1a Series circuits

You will investigate changes in voltage and current in a series circuit.

Apparatus

battery

three lamps

ammeter

leads

voltmeter

Method

1. Assemble each of these two circuits in turn. Note the readings on the meters each time.

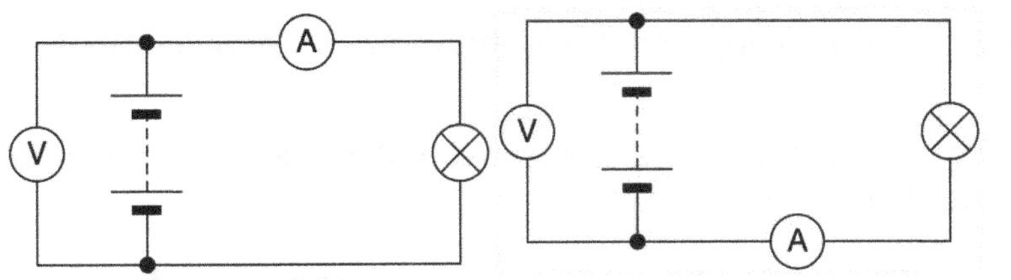

2. Assemble these three circuits in turn. Note the readings on the meters each time.

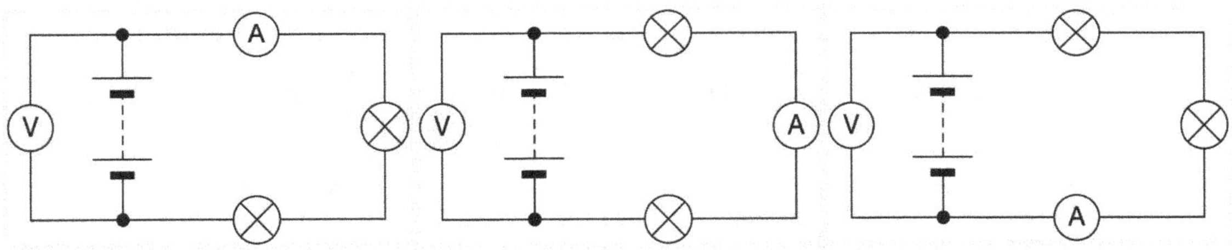

3. Now use the ammeter to investigate what happens as current passes through three lamps in series with a battery.

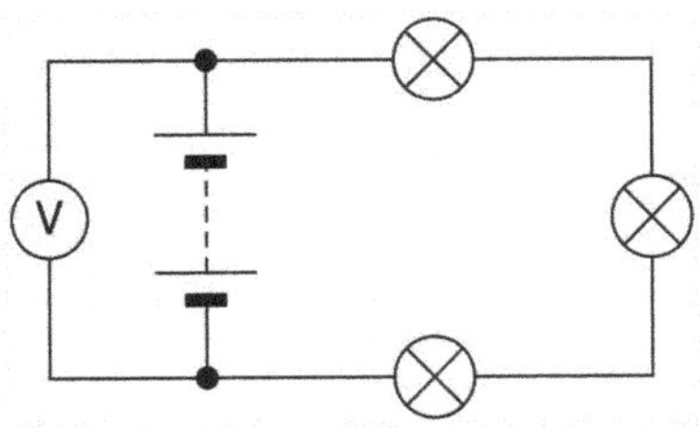

Analyse and interpret data

1. What happens to the voltage and current as you increase the number of lamps?

P2c.1b Parallel circuits

You will investigate level of current in a parallel circuit.

Apparatus

battery

ammeter

22 Ω resistor, 10 Ω resistor

leads

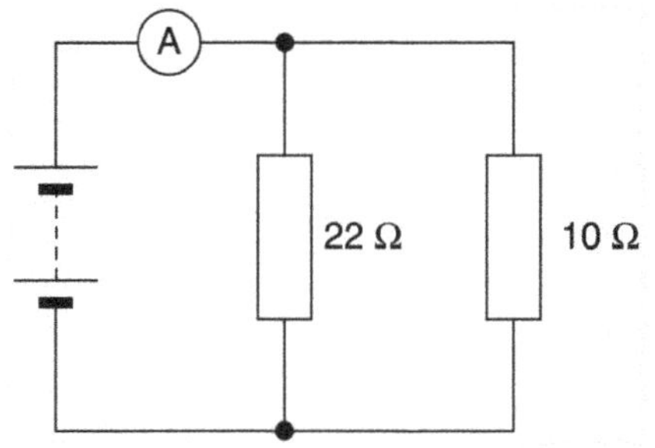

Method

1. Assemble the circuit.

2. The ammeter shows the current going from the battery into the parallel resistors. Record its value – it should be about 0.4 A.

3. Assemble these circuits to measure the current in each of the resistors separately. Record their values. They should be different and less than 0.4 A.

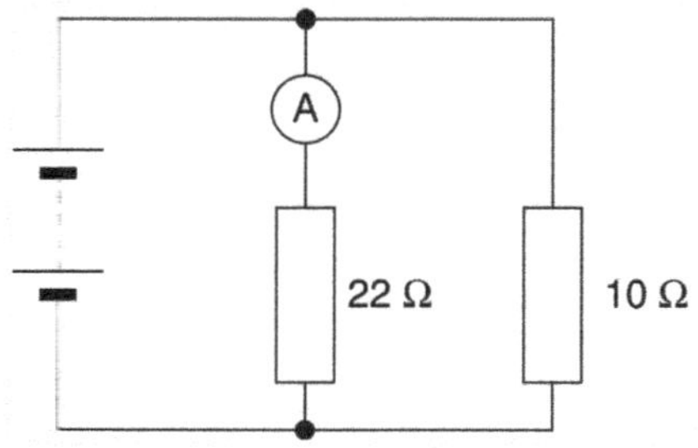

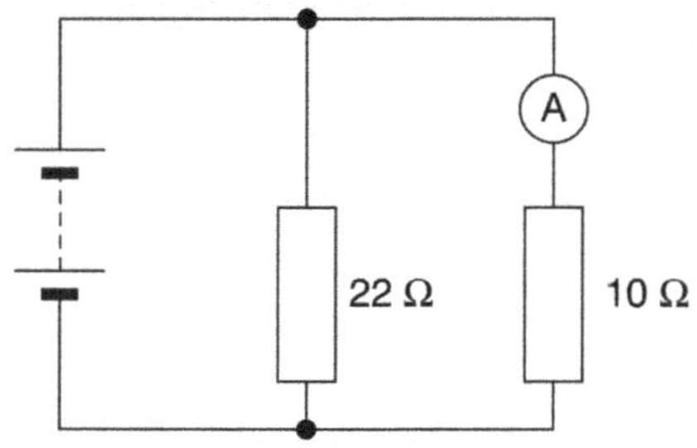

4. Finally, use the circuit below to measure the current going from the resistors back into the battery. Is it what you expect it to be?

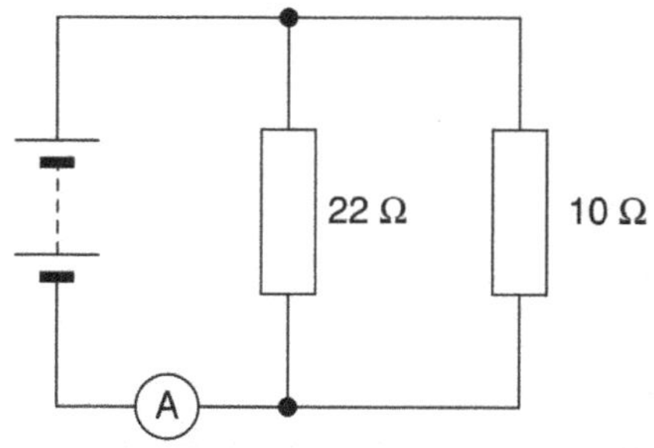

Analyse and interpret data

1. What can you say about the sum of all current in all the branches of the circuit in comparison to the current measured coming out of the power source?

Activity P2c.2 How changing resistance affects current

Specification reference: P2c.2.8, P2c.2.11

Learning objectives

- Understand how the current in a series circuit depends on the voltage and the components in the circuit
- Understand the circuit operation for an ohmic conductor (e.g. resistance wire)
- Understand the behaviour of a number of non-ohmic conductors
- Describe the variation of resistance of light-dependent resistors with illumination and thermistors with current

Learning outcomes

- Define resistance
- Describe experiments to find *V–I* graphs for various components
- Recognise non-ohmic graphs
- Summarise the behaviour of non-ohmic components
- Describe experiments to demonstrate the behaviour of non-ohmic components
- Recognise the temperature–resistance graph (thermistor) and intensity–resistance graph (LDR) as characteristic of the component

Common misconceptions

Students often find it difficult to relate the shape of a current–voltage graph to the varying resistance of a component. They also find the negative portions of these graphs confusing.

Voltmeters and ammeters look very similar so students often confuse current with voltage.

Resources

Student Book pages 97–100

Worksheet P2c.2a Voltage and current in a filament lamp

Worksheet P2c.2b Voltage and current in a thermistor

Resources for class demonstrations and practicals (see Technician's notes, following)

Approach

Introduce the activity

Connect a thermistor in series with a battery and an ammeter as shown.

Place the thermistor in a test tube to keep it dry. Place the test tube in a beaker and pour freshly boiled water into the beaker to raise the temperature of the thermistor. Use class questioning to establish that the resistance of the thermistor has fallen.

Connect an LDR in series with a battery, ammeter and LED as shown in the diagram.

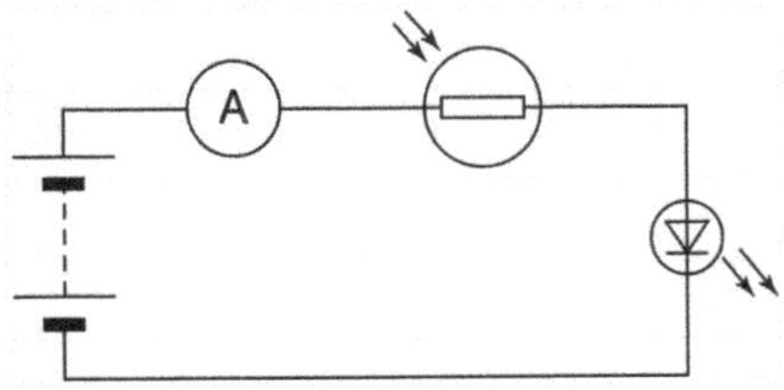

Shine the torch on the LDR. Use class questioning to establish that the LDR resistance has gone down. Then draw their attention to the LED. Turn it round in the circuit. Use class questioning to establish that it now has a very large resistance (the current should be essentially zero).

Develop the activity

Discuss the content on page 97 of the Student Book. At this point the students can carry out the practical on Worksheet P2c.2a. Students use the variable resistor to set the current in the lamp to several different values, noting the voltage each time. They will then use their results to plot a graph to show how the current depends on the voltage.

It is a good idea to check each group's circuit before you let them take readings. Some students may need assistance with setting up the axes of their graph.

Discuss 'The thermistor and the LDR' on page 99 of the Student Book. Students can then carry out the practical on Worksheet P2c.2b. Students might be interested to know that an understanding of semiconductors is fundamental to modern communications and information technology.

Students should then tackle the questions and the Developing investigative skills feature on page 100 of the Student Book in pairs.

Students might relish the opportunity to assemble and test the diode circuit shown below. An LED could act as the chip.

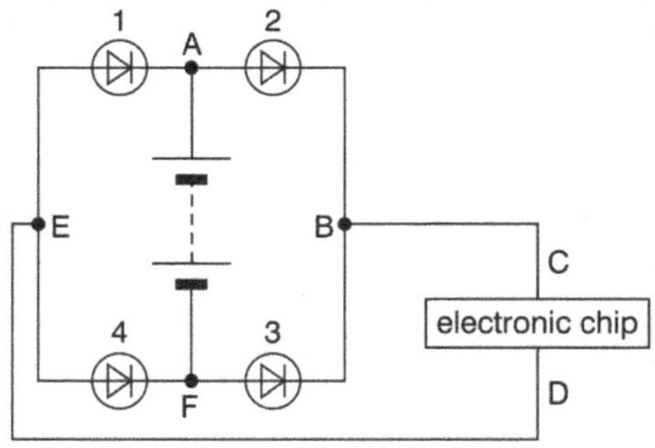

Bring the activity together

Ask the class to suggest why shining light on an LDR reduces its resistance (it increases the number of free electrons). This idea will be explored further in the next activity.

Technician's notes

Be sure to check the latest safety notes on these resources before proceeding.

The following resources are needed for the class demonstrations P2c.2.

Introduction

thermistor
ammeter (0–20 mA)
6V battery
hot water from the tap
test tube to hold thermistor
beaker for water
six leads
LDR
torch
LED (with inbuilt 100 Ω safety resistor)

A 300 Ω NTC disc thermistor should be fine with a 3 V battery pack. The thermistor will be kept dry in the test tube – its leads will therefore need to be quite long. The second starter suggestion has featured in a previous lesson.

The LED is best viewed in subdued lighting.

Main part of the activity

four diodes
LED with built-in 100 Ω safety resistor
6 V battery
ten leads

The following resources are needed for the class practical P2c.2a, per group:

3 V battery pack
ammeter (0–1 A)
voltmeter (0–10 V)
filament lamp, diode, LED
variable resistor (0–100 Ω)
4 mm connecting leads
graph paper

The lamp rating needs to match the battery voltage.

3.5 V MES bulbs are easy to match with a 3 V battery of two 1.5 V cells.

The resistor needs to be about the same as the working resistance of the lamp: 12 Ω is fine for a 3.5 V bulb which draws 0.3 A.

The variable resistor needs to go from 0 to about ten times the working resistance of the lamp: 100 Ω is fine for a 3.5 V bulb.

The following resources are needed for the class practical P2c.2b, per group:

3 V battery pack
ammeter (0–1 A)
voltmeter (0–10 V)
thermistor
resistor to limit current
beakers containing water at about 50 °C, room temperature and with ice
thermometer
4 mm connecting leads
graph paper

A suitable range for a class experiment such as this might be around 20–25 °C either side of room temperature. This allows hot water to be the heat source to warm the thermistors up (and ice will cool them down) and it also reduces the risk of harm to students if the temperatures are kept moderate.

Answers

Page 100

1. In brighter light, the resistance of the LDR decreases, so the current in the circuit will increase. The ammeter reading will therefore increase in brighter light.

2. A diode has a low resistance if the current moves in one direction and a very high resistance in the other. A waterfall allows water to move easily in one direction, but the water cannot go 'backwards' up the waterfall. In this way the analogy is reasonable.

3. Increasing the temperature reduces resistance.

Developing investigative skills, pages 100–101

1. To make sure the water is all at the same temperature. The thermometer actually records the temperature of the water, so if the student is going to use this value as the temperature of the thermistor, she needs to ensure all the water is at the same temperature.

2. a) The range of the thermometer is 60 °C.

   b) When it records 40 °C the true value will be between 38 °C and 42 °C.

3. Graph.

4. As the temperature increases, the resistance decreases.

5. Similarly to question 1, the thermometer is actually measuring the temperature of the water. If the student wishes to use this value as the temperature of the thermistor, then she must allow time for the energy transfer between the thermistor and the water to reach equilibrium.

6. As the temperature increases the resistance decreases, so at higher temperatures it becomes easier to achieve a net transfer of electrons. Increasing the temperature would increase the average kinetic energy of the particles, but this would be in a random manner and would particularly apply to the large ions in the structure (which is why the resistance of a filament lamp increases as the temperature increases). In this case, the additional energy makes more electrons available to conduct which overwhelms the increased resistance due to increased particle vibrations leading to a decrease in resistance overall.

P2c.2a Voltage and current in a filament lamp

In this practical, you will study the voltage and current in a filament lamp.

Apparatus

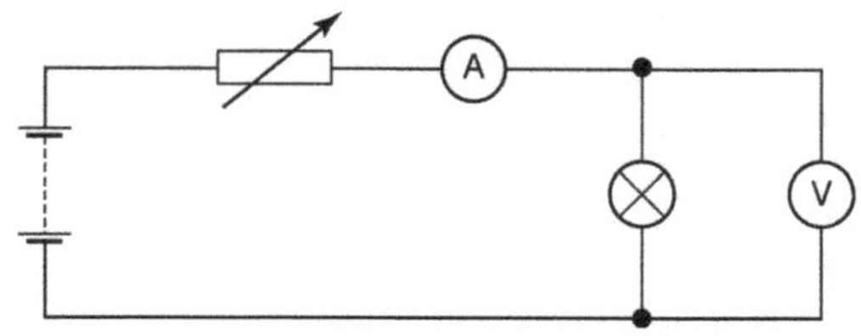

battery

ammeter

voltmeter

leads

variable resistor

filament lamp

graph paper

Method

1. Connect the variable resistor, ammeter and lamp in series with the battery.
2. Alter the variable resistor. If all is well, the ammeter reading should change from about 0.03 A to about 0.30 A.
3. Connect the voltmeter in parallel with the fixed resistor. It should have a maximum voltage reading of 3 V.
4. Alter the variable resistor. If all is well, the voltmeter reading should go up and down as the ammeter reading goes up and down.
5. Measure the current in the lamp for ten different values of voltage across it. Try to cover the whole range of values evenly. For each setting of the variable resistor, check that turning the lamp round the other way in the circuit makes no difference to the readings.
6. Enter your results in a table like this.

Potential difference (V)	**Current (A)**

Analyse and interpret data

1. Use your results to plot a current–potential difference graph.
2. What do you notice about the shape of your graph?

P2c.2b Voltage and current in a thermistor

In this practical, you will investigate the link between voltage, current and temperature.

Apparatus

battery

ammeter

voltmeter

leads

thermistor

resistor

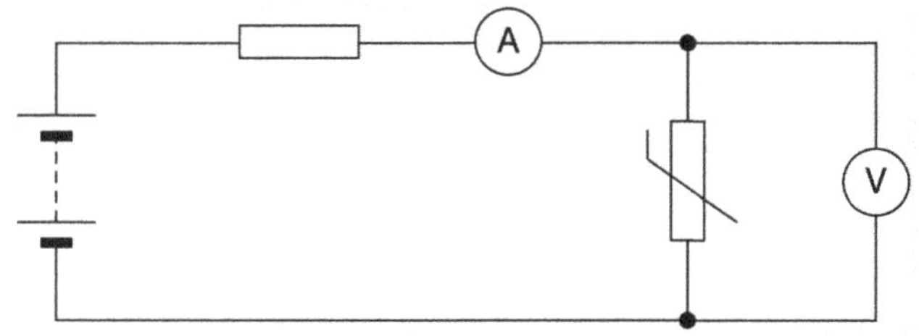

beaker containing water at room temperature, about 50 °C and just above freezing

graph paper

Method

1. Connect the ammeter, thermistor and resistor in series with the battery.
2. Connect the voltmeter in parallel with the thermistor.
3. Put the thermistor in one of the beakers containing water. Leave it for a couple of minutes so that its temperature is the same as the water.
4. Measure the temperature of the water and the current through the thermistor and voltage across the thermistor.
5. Enter your results in a table like this:

Voltage (V)	**Current (A)**	**Temperature (°C)**

Analyse and interpret data

1. Use your results to plot a graph of voltage against current.
2. What do you notice about how the voltage and current vary with temperature?

Activity P2c.3 Voltage, current and resistance

Specification reference: P2c.2.9, P2c.2.10, P2c.2.12, P2c.2.13

Learning objectives

- Describe how current varies with voltage in wires, resistors, lamps and diodes
- Describe the effect of changing resistance on the current in a circuit
- Learn the relationship between voltage, current and resistance: voltage = current × resistance $V = I \times R$

Learning outcomes

- Use the equation $V = I \times R$ correctly
- Be able to explain the flow of current in metallic conductors

Common misconceptions

As soon as physics becomes quantitative it becomes much more difficult for many students. Voltmeters and ammeters look very similar so students often confuse current with voltage.

Some uncertainty at this point is perfectly normal. As long as the correct language is used and misconceptions are corrected every time they crop up, students should get things straight in their heads by the end of the topic.

Students often find it difficult to relate the shape of a current–voltage graph to the varying resistance of a component. They also find the negative portions of these graphs confusing.

Resources

Student Book pages 101–107

Worksheet P2c.3a Voltage–current graphs

Resources for class practical (see Technician's notes, following)

Approach

Introduce the activity

Remind students about the shape of a *V–I* graph for a conductor that they met in the last activity. Ask them to tell you what quantity is shown by the gradient of a *V–I* graph (resistance). Explain that in this activity the relationship between *V, I* and *R* is going to be explored in more detail.

Develop the activity

Discuss 'Voltage, current and resistance' on page 101 of the Student Book. Students can then do the practical on Worksheet P2c.3a. Students will assemble a circuit to measure the current in the resistor at a variety of voltages across it. They will use their results to calculate the resistance of the resistor and draw a graph to show how the current depends on the voltage.

Students may need help with evaluating their results because they do not fully appreciate how the uncertainty of their values for voltage and current affect their calculated values for resistance. This is especially true for low values of current where errors may easily reach 25%. The graph shows this clearly.

Students should tackle the worked examples of page 103 in the Student Book without the given solutions. Then compare students' solutions with the given solutions and discuss any differences.

Students can then tackle the questions on page 103 of the Student Book. You should encourage students to write down all of their working when they do the calculations of current, voltage and resistance. It is very easy for them to get lost if they just use their calculators directly.

Discuss 'Current is the rate of flow of charge' on page 104 of the Student Book (note that the content continues on page 105). Refer back to the question you posed at the end of the previous activity about why shining light on an LDR reduces its resistance. Students should now see why the answer to the question is that 'light increases the flow of electrons'. Students can also be introduced to the idea of combining resistors in series on page 103. A qualitative and quantitative understanding is useful here.

Students can now study the Developing investigative skills feature on page 102 of the Student Book in pairs. Students should familiarise themselves with the conservation of current at a junction on page 105 then tackle the questions on page 105.

Discuss 'Current in metallic conductors' on page 106 of the Student Book. Students can then tackle the question on page 106 of the Student Book.

Ask students to make notes on the Science in context feature on page 106 of the Student Book.

Bring the activity together

Ask students to write down two things they think they did well in this activity and one thing that they think they could have done better. Discuss the areas in which they think they could improve.

Technician's notes

Be sure to check the latest safety notes on these resources before proceeding.

The following resources are needed for the class practical P2c.3a, per group:

3 V battery pack
ammeter (0–1 A)
voltmeter (0–10 V)
12 Ω resistor (1 W rating at least)
variable resistor (0–100 Ω)
4 mm connecting leads

For a filament lamp, use a ray box if available, or alternatively any low power bulb where there will be a definite rise in temperature as it moves towards full operating voltage.
Digital meters are useful if they are available.

Answers

Page 103

1. 10 V.

2. 6.0 Ω.

3. The current would rise ever more slowly as the potential difference increased. This is because increasing the current makes the wire hotter, increasing the resistance.

Developing investigative skills, page 102

1. The method involves a comparison between the voltmeter reading and the ammeter reading. These will change in proportion (Ohm's law) so the absolute values do not matter, the ratio will remain constant.

2. A zero error is where the meter has an incorrect reading when it should in fact read zero. To check the ammeter and voltmeter, the student should connect together the terminals of each meter – if the reading is not zero then the scale on the meter will need to be adjusted.

3. Graph drawn – best fit line should be a straight line through the origin.

4. Resistance is given by the gradient of the line drawn on the graph. The triangle used for the calculation should be clear on the graph and should be of a good size – generally a triangle spreading over at least half the graph line would be taken as the minimum acceptable.

5. Calculating the resistance from each pair of readings and then taking a mean would give equal importance to the particular values measured. When drawing the best fit line on the graph you can make a finer judgement about the value of each data point, weighting the line towards the more general pattern.

6. Heating effects can be reduced by actions such as carrying out the experiment at lower supply voltages and only completing the circuit for short periods of time when measurements are actually being made.

Page 105

1. a) 15 C.

   b) 20 C.

   c) 92 C.

   d) 45 C.

2. 9000 C.

3. 120 s.

Page 106

1. They flow in opposite directions.

P2c.3a Voltage–current graphs

In this practical, you will study how the current depends on the voltage.

Apparatus

battery
voltmeter
ammeter
fixed resistor
variable resistor
leads

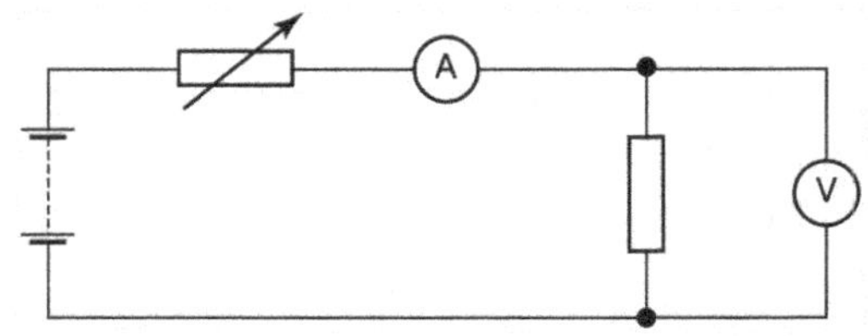

Method

1. Connect the variable resistor, ammeter and fixed resistor in series with the battery.

2. Alter the variable resistor. If all is well, the ammeter reading should change from about 0.03 A to about 0.25 A.

3. Connect the voltmeter in parallel with the fixed resistor.

4. Alter the variable resistor. If all is well, the voltmeter reading should go up and down as the ammeter reading goes up and down.

5. Measure the current in the fixed resistor for ten different values of voltage across it. Try to cover the whole range of values evenly.

6. Enter your results in a table like this.

Voltage (V)	Current (A)	Resistance (Ω)

Analyse and interpret data

1. Use your results to plot a current–voltage graph. Plot voltage (*V*) on the *y*-axis and current (*I*) on the *x*-axis.

2. Draw a line of best fit though the points.

3. Use your line of best fit to read off a single pair of values for current *I* and voltage *V*.

4. Use your values for *I* and *V* to calculate a value for the resistance $R = \frac{V}{I}$.

5. What do you notice about the graph?

6. Find the gradient of the graph.

7. Use the rule $\text{resistance} = \frac{\text{voltage}}{\text{current}}$ to fill in the last column of the table.

8. Do the results agree with the idea of resistance being the same for all values of voltage? Use your data to justify your answer.

Activity P2c.4 Voltage, energy and charge

Specification reference: P2c.2.14, P2c.2.15, P2c.2.16, P2c.2.17, P2c.2.18, P2c.2.19, P2c.2.20, P2c.2.21

Learning objectives

- Know that current is the rate of flow of charge
- Know and use the relationship between charge, current and time: charge = current × time
  $Q = I \times t$
- Know that current in solid conductors is the flow of negatively charged electrons
- Know that voltage is the energy transferred per unit charge passed, meaning that the volt is one joule per coulomb
- Know and use the relationship between voltage, energy and charge: energy transferred = charge × voltage
  $E = Q \times V$

Learning outcomes

- Be able to use $Q = I \times t$
- Be able to define the volt
- Be able to use $E = Q \times V$
- Apply ideas about conservation of charge to explain observations from measurements of current in simple circuits

Common misconceptions

There is confusion caused by the difference between the direction of conventional current (arbitrarily chosen by Benjamin Franklin) and that of electron flow in a metallic conductor.

Students want something to be 'used up' as it goes around a circuit. This gets in the way of their accepting the idea of conservation of current in a series circuit. It helps to emphasise from the start that the current carries energy which is transferred on its way round the circuit.

Resources

Student Book pages 107–110

Worksheet P2c.4a Voltage

Resources for class demonstration and practical (see Technician's notes, following)

Resources for class demonstration in the main part of the activity: Pile of A5 size pieces of paper.

Approach

Introduce the activity
Set up a series circuit on the bench with a resistor, LED and battery, but no meters. Discuss how the circulating free electron model can explain the transfer of energy from the battery to the resistor and the LED, each of which transfers energy to the environment as heat or light. Use class questioning to establish that if the wires are full of free electrons, current into a component must equal current out if current is related to speed of electron flow.

Develop the lesson

Discuss 'Voltage, energy and charge' and 'Energy transferred, charge and voltage' on pages 107–108 of the Student Book. Make sure that students can see how the equation $E = Q \times V$ and $V = \frac{E}{Q}$ are derived.

Look at the worked example to consolidate the learning. Move on to discuss how electricity is measured (page 109 of the Student Book). At this point the students can carry out the practical. Students will use the voltmeter to investigate the voltage across the components in a series circuit of three lamps and a battery.

It is worthwhile emphasising at this stage that the voltage across a component is a measure of how much energy is transferred by the current, so it can get transferred from the electrons to the components on its way round the circuit.

Revisit the model for electric current used in Activity P2c.1

Explain that the students are going to model an electric circuit. It will not be a perfect model, but it will include a number of the important points. Choose one student to be the 'cell' and have them sit on a chair at the front of the room. Give him/her a pile of pieces of paper (quite a lot – pages of used photocopies, or an old telephone directory, cut up into pieces of roughly A5 size). The paper is the energy stored in the 'cell'.

Choose another student to be a 'bulb' and have them sit at a convenient place in the room. Ask students what will happen next – nothing, because there is no circuit. Identify a path round the room that will be the circuit. What will happen next? Nothing, because there is no way to get the 'energy' (the paper) from the 'cell' to the 'bulb'.

Choose another student to be an electron and walk them very slowly through the process of collecting 'energy' from the cell and carrying it round to the 'bulb'. Transfer the energy to the 'bulb' – the bulb student rips the paper in two (representing heat and light) and the bulb 'transfers' the energy to the surroundings. Keep expanding the model by adding more electrons.

Now add a second 'cell' in series with the first. Each electron now picks up two pieces of paper – one from each 'cell'. What difference will there be at the 'bulb'? It is 'brighter' because it receives more energy. Now add a second 'bulb in series' with the first – what difference will that make? The bulbs will be 'dimmer' because the 'electrons' give one piece of paper to each bulb. And so on …

Next model two bulbs connected in parallel: 'electrons' have to go one way or the other, so the 'current' splits. Add students to represent 'ammeters' around the circuit and ask them to count the number of student electrons going past. The rules for currents in series and parallel circuits will be apparent.

Add students to represent 'voltmeters' placed opposite 'bulbs' and 'cells' and ask them to count the number of pieces of paper being transferred. The rules for voltage will not appear as clearly as the current values, but they will if you divide the number of pieces of paper by the number of students – leading to the idea of energy transferred per unit charge.

The extension box on page 110 contains an alternative model for the flow of electric charge. All students should tackle the questions on page 110 of the Student Book in pairs.

Bring the activity together

Write the words 'voltage', 'current', 'resistance', 'energy', 'charge', 'electrons' on the board. Ask students to work with a partner to create a spider diagram using these words. After a short period of time, take feedback. Allow students to add to their spider diagram as necessary.

Technician's notes

Be sure to check the latest safety notes on these resources before proceeding.

The following resources are needed for the class demonstration P2c.4.

Introduction

battery
LED
100 Ω resistor
4 mm leads

The following resources are needed for the class practical P2c.4a, per group:

battery
three lamps
voltmeter (0–5 V)
seven 4 mm leads

Answers

Page 110 (top)

1. The battery or power source.
2. They transfer the energy to the components in the circuit.
3. The difference in voltage between the two points.
4. Yes, voltage is the difference in energy of a coulomb of charge between two points in a circuit.
5. The electrons at the point with the higher voltage will have more energy than the electrons at the point with the lower voltage.

Page 110 (bottom)

1. Students' own answers. One possible answer is that the current is the amount of water flowing, the potential difference is height the water is pumped up. The ammeter measures the amount of water flowing past the water wheel, the voltmeter measures the height difference. A series circuit would have two or water wheels connected one after the other, a parallel circuit would have two or more water wheels connected side by side.
2. Students' own answers. One possible answer is that it is a good model, but does not model the way current is affected by resistance in series and parallel circuits. The amount of water flowing relies on how much water the pump is pumping.

P2c.4a Voltage

In this practical, you will assemble a variety of circuits and use them to find the voltages across different bulbs and the connection between these.

Apparatus

battery

three lamps

leads

voltmeter

Method

1. Assemble each of these circuits in turn. Note the readings on the voltmeter each time.

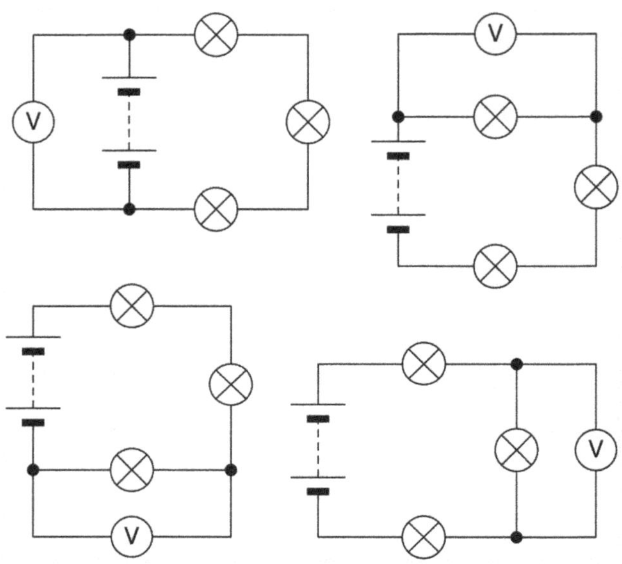

Analyse and interpret data

Use your findings to answer these questions:

1. Three lamps are connected in series to a 6 V battery. What is the voltage across each lamp? Which lamp carries the most current?

2. The current in a single lamp is 0.20 A when connected across a 1.5 V cell. Four of these lamps are connected in series with a 6.0 V battery. Explain why the current in each lamp is still 0.20 A.

Activity P2c.5 Consolidation and summary

Learning objectives

- To review the learning points of the topic
- To test understanding through answering questions

Learning outcomes

- Be familiar with the knowledge and understanding summarised in the End of topic checklist
- Be able to apply this knowledge and understanding by answering End of topic questions

Resources

Student Book pages 111–116

Approach

Introduce the activity

Ask students to work with a partner to make a list of key words from this topic. They could then work together to produce a spider diagram showing how the different concepts are linked. They could compare their list with the list of key words given on page 111 of the Student Book. Discuss the checklist given on pages 111–112 and use questioning to see how much of the content students are comfortable with.

Develop the lesson

Ask students to work individually to work through the End of topic questions on pages 113–116 of the Student Book without reference to the text. As they work, walk around the classroom observing their answers and questioning them as necessary to find out which questions are causing difficulties.

Bring the lesson together

After a set period, ask the students to stop working and discuss any areas of difficulty you observed as you walked round the class. Students should complete any unanswered questions for homework, but you should stress that they should attempt the questions without reference to the text to see how much content they have retained.

Answers

End of topic questions mark scheme

Question	Correct answer	Marks
1 a)	A_1 reading 0.2 A A_2 reading 0.2 A. The current is always the same at any point in a series circuit	1 1
1 b)	A_4 reading 0.30 A A_5 reading 0.15 A. The ammeter A_6 has been placed on one branch of the parallel circuit. Ammeter A_5 is on the other branch. As the lamps are identical the current flowing through them must be the same. Ammeter A_4 gives the current before it 'splits' in half as it flows through the two parallel branches	1 1
2	reading on V_1 = 6 V. The voltage across the battery is 9 V. This must equal the total voltage in the circuit. Assuming there is no loss along the copper wiring the voltage across the lamp must be 9 − 3 = 6 V	1 1
3 a)	$I = Q/t$, I = 10/30 = 0.33 A	1 1
3 b) i)	10 C	1 1
3 b) ii)	36 000 C	1 1
4 a)	24 V	1 1
4 b)	20 V	1 1
4 c)	0.12 A	1 1
4 d)	23 A	1 1
4 e)	60 Ω	1 1
4 f)	23 Ω	1 1
5 a)	the battery	1
5 b)	the voltage will increase, and so will the current	1
5 c)	it will decrease	2

Question	Correct answer	Marks
5 d)	1.5 V	1 1
6 a)	a line graph with voltage on the *x*-axis and current on the *y*-axis	1 1
6 b)	voltage	1
6 c)	they are consistent but he only took one set of readings	1 1
6 d)	he could have repeated them he could have recorded more values	1 1
6 e)	correctly plotted graph resistance = 3.5 Ω	2 1
7	row 1: 0.3 (V) row 2: 30 (Ω) row 3: 6 (V) row 4: 4 (Ω) row 5: 13 (12.97) (A)	1 1 1 1 1
8 a)	$A_1 = A_2 = 0.3$ A $V_1 = V_2 = 1.5$ V. The current through each lamp would be the same but the voltage across each lamp would be different	2 1 2 1
8 b)	$A_1 = A_2 = 0.15$ A $V_1 = V_2 = 3.0$ V. The voltage across each lamp would be the same, but the current through each lamp would be different	2 1 2 1
9 a)	series	1
9 b)	parallel	1
9 c)	parallel	1
9 d)	series	1
9 e)	series	1
9 f)	parallel	1
9 g)	parallel	1
9 h)	series	1
10 a)	positions of ammeters and voltmeters should be reversed	1 1

Question	Correct answer	Marks
10 b)	voltage across variable resistor (V), voltage across lamp (V) current through main circuit (A), current through lamp (A)	1 1
11	First, calculate the total resistance of the three resistors: Write down the formula: $V = I \times R$ Rearrange the formula: $R = V \div I$ Substitute the values for V and I: $R = 24 \div 6$ Work out the answer and write down the units: $R = 4\ \Omega$ Total resistance = $4\ \Omega = 1 + X + X$ Rearrange the equation: $2X = 4 - 1 = 3$ $X = 3 \div 2 = 1.5\ \Omega$ the unknown resistors each have a value of $1.5\ \Omega$	 1 1 1 1
	Total:	73

P2d Electric charge

Introduction

In this topic students are introduced to the concept of electric charge. They learn about insulators and conductors. This topic introduces electrostatics and how it is used.

Links to other topics

Section	Essential background knowledge	Useful links
2 Electricity	P2b Mains electricity P2c Energy and voltage in circuits	

Topic overview

P2d.1	**Charging by friction – attraction and repulsion** In this activity students experience experiments to demonstrate simple electrostatic effects. They learn how to use of the word 'charge' and why there is the need for two types. They learn an explanation for charge in terms of electrons.
P2d.2	**Uses and hazards of electrostatic charges** This activity introduces devices which make use of electrostatic phenomena. It also introduces hazards associated with electrostatic charges and the measures that can be taken to reduce these risks.
P2d.3	**Consolidation and summary** This activity provides an opportunity for a quick recap on the ideas encountered in the topic as well as time for the students to answer the End of topic questions in the Student Book.

Activity P2d.1 Charging by friction – attraction and repulsion

Specification reference: P2d.2.22P, P2d.2.23P, P2d.2.24P, P2d.2.25P, P2d.2.26P

Learning objectives

- Identify materials which are electrical conductors and electrical insulators
- Use the word 'charge' and understand the need for two types
- Explain electrostatic charge in terms of loss and gain of electrons
- Explain electrostatic phenomena in terms of movement of electrons

Learning outcomes

- Conduct experiments to demonstrate simple electrostatic effects
- Describe how to charge an insulator by friction and explain why an insulator is needed
- Explain why there must be two types of charge, describing simple experiments which support this idea
- Be able to link positive and negative charges correctly to the movement of electrons.

Common misconceptions

Students need to appreciate the particular use of the word 'charge' to explain static electricity. For many, charging has to do with energy transfer, as in charging a mobile phone. The idea of 'charge' as an attribute of matter similar to mass which is responsible for forces is a difficult one to learn when the word already means something else.

Resources

Student Book pages 117–121

Worksheet P2d.1a Friction charging

Resources for demonstration (see Technician's notes, following)

Approach

Introduce the activity

You could use a Van de Graaff generator to set up some large sparks. Explain that the sparks are caused by rapid heating of the air as electrons jump from one sphere to the other.

Develop the activity

Discuss 'Conductors and insulators' on page 117 of the Student book. Move on to discuss charging by friction. At this point the students should carry out the practical on Worksheet P2d.1a. Students will dry one end of each polymer strip, then pull that end through the cloth to become charged.

One strip will then be balanced on the evaporating dish so that forces of attraction or repulsion between it and the other strips can be observed.

Discuss 'Charge as the loss or gain of electrons' on page 119 of the Student Book. Students should attempt the question on page 121 in pairs. The fact that charged objects attract neutral ones can form a barrier to understanding, particularly in practical work. It is best if students find this out early on, but there is no need at this stage for them to understand why it happens in terms of polarisation.

Demonstrate charging balloons as follows. Tell students that they need to note down any observations as you do the demonstration.

Each balloon is inflated and then grasped at the top in one hand.

The cloth is rubbed briskly over the surface of the balloon not being grasped. The balloon is then suspended from the clamp stand by the thread.

The other balloon is similarly charged and suspended.

They are then brought towards each other by moving the stands together. If all is well, the balloons should repel each other, suggesting a rule: similarly charged objects repel.

Charge each strip in turn and hold close to one of the balloons.

The attraction to one of the strips suggests that there is more than one type of charge, leading to another rule: differently charged objects attract.

The demonstration works best if the equipment is dry. Liberal use of the hair dryer may help.

Move on to summarise the students' observations and link these with 'Attraction and repulsion' on page 120 of the Student Book. Students need to learn that electrons have negative charge so removing them from a neutral atom leaves the atom with positive charge.

Students should tackle the questions on page 121 in pairs.

Students are now in a position to tackle the worksheet.

Bring the activity together

Use a Van de Graaff generator to charge a metallised plastic ball on a nylon thread. Note the reaction of the ball to strips of charged cellulose acetate and charged polythene. Ask the class to deduce the charge on the dome of the generator.

Charge a balloon by rubbing it on a student's hair. Suspend the balloon from a thread. Get the class to talk you through how to establish the charge left on the student's hair. [Use cellulose acetate and polythene strips to find the charge on the balloon.]

Technician's notes

Be sure to check the latest safety notes on these resources before proceeding.

The following resources are needed for the class demonstrations P2d.1:

Introduction

Van de Graaff generator
metal sphere on insulating handle
lead to connect sphere to base of generator
rubber balloon

It is important for the safety of the demonstrator that the base of the generator is properly earthed.

No student or member of staff should have electrical item in their pockets or anywhere on them, or have a cochlear implant or other hearing aid fitted or have a heart condition of any kind, or have a pacemaker implanted.

The Van de Graaff generator should be at least 2 m away from any electrical equipment, e.g. computer, mobile phones and calculators.

Main part of the activity

pair of rubber balloons
thread to suspend the balloons

pair of clamp stands to suspend balloons
cellulose acetate strip
polythene strip
woollen cloth
acetate bar
cloth
polythene bar
small pieces of paper

The following resources needed for class practical P2d.1a, per group:

woollen cloth	one strip of wood
two strips of cellulose acetate	hair dryer
two strips of polythene	small evaporating dish

The experiment works best if the materials are dry.

Answers

Page 120

1. You are the insulator. You will have rubbed electrons either onto or off yourself, perhaps by sliding your feet over a carpet. The metal handrail is a conductor – when you touch it, it allows the electrons to move to restore the balance and this is what you feel as a shock.

Developing investigative skills, page 120

1. Rub a rod with the dry cloth and hold it near small pieces of paper. If the pieces of paper move towards the rod, the rod is charged. Also accept: hold the rod near a stream of water from a tap, see if the water moves towards the rod.

2. Suspend two of the rods each from a retort stand using a length of nylon string. Make sure the rods are far from any metal supports. Rub each of the rods with the dry cloth. Move the two stands closer together so that the rods are almost touching. If the rods have the same charge (e.g. both positive) they will repel each other. If they have unlike charges the rods will attract.

3. Dry air is an insulator so electrostatically charged objects remain charged. If the atmosphere is humid, the charged objects lose their charge easily because water conducts electricity very well.

4. The metal rod, because the metal conducts any electrons transferred to it from the cloth through your hand to conduct to the ground: it does not stay charged.

Page 121

1. Electrons are extremely small and negatively charged. Electrons are also around the outside of atoms. These key ideas make it much easier for electrons to be moved about (than the positively charged parts of the atom) to account for all the electric effects we know.

2. The electron clouds in your feet repel the electron clouds in the floor. The force of repulsion is easily enough to stop you falling through.

3. a) Polythene rod should be negatively charged and cloth positively charged. Perspex rod should be positively charged and cloth negatively charged.

b) They will be equal and opposite.

P2d.1a Friction charging

In this activity, you will investigate the effect of charging different materials.

Apparatus

polythene strips

cellulose acetate strips

wooden strip

woollen cloth

evaporating dish

hair dryer

Method

1. Place the evaporating dish upside down on the table.
2. Balance the wooden strip on the dish so that it can rotate freely.
3. Hold the polythene strip at one end. Dry the other end with the hair dryer.
4. Pull the dry end of the strip briskly through the cloth. Do it several times. This should give the end a **negative** charge.
5. Hold the charged polythene close to the **neutral** wooden strip. There should be a force pulling them together.
6. Place the charged strip on the dish so that it can rotate freely. Make sure you know which end is charged.
7. Charge the other polythene strip. Hold it close to the charged end of the strip on the dish. If all is well, the strips should repel each other.
8. Charge the cellulose acetate strip and hold it close to the charged polythene strip. They should attract each other.
9. Replace the polythene strip on the dish with the charged cellulose acetate strip.
10. Charge the second cellulose acetate strip and hold it close to the one on the dish. They should repel each other, showing that they have **positive** charge.
11. Place the neutral wooden strip on the dish. Check that it is attracted to the charged cellulose acetate strip.

Activity P2d.2 Uses and hazards of electrostatic charges

Specification reference: P2d.2.27P, P2d.2.28P

Learning objectives

- Explain hazards associated with electrostatic charges (e.g. fuelling aircraft)
- Discuss devices which make use of electrostatic phenomena (e.g. photocopiers)

Learning outcomes

- Be able to describe simple examples where electrostatic effects pose a potential hazard, e.g. in refuelling
- Be able to describe measures taken to reduce the associated risks, e.g. use of appropriate materials
- Be able to describe the operation of a device, such as a photocopier or a precipitator in a chimney, that use electrostatic effects

Common misconceptions

Students may not realise that electrostatics can be useful.

Resources

Student Book pages 122–125

Approach

Introduce the activity

Ask students what they know about how photocopiers and laser printers work. The answer is electrostatics. Explain that in this activity students will learn more about the hazards and uses of electrostatics.

Develop the activity

Ask students whether they have experienced a small shock when combing their hair, or pulling a jumper over their head. Discuss techniques that are used to protect sensitive equipment from such shocks, such as antistatic wrist straps like the one shown in Figure 2.50 on page 122 of the Student Book.

Now move on to discuss uses of electrostatics. Use the text on page 122 of the Student Book to guide the discussion. Students should then study the Science in context feature on page 123 before tackling the questions on page 125 in pairs.

Bring the activity together

In pairs, students should write down two things that they have learnt in this lesson and one thing that they would like to know more about. They can research the thing that they would like to know more about as part of their homework.

Answers

Page 125

1. Any suitable example such as combing your hair, touching a door knob or a car.

2.

Situation	Danger	Prevention
(Refuelling a car)	(Sparks could form as charge builds up causing an explosion)	(Tyres contain graphite – a conductor that transfers electrons to earth)
Making sensitive electronic equipment	High voltages can damage (low voltage) electronic equipment	Worker use earth straps to transfer electrons to the ground and prevent them becoming charged
Refuelling an aeroplane full of passengers	High voltages can cause sparks which can ignite fuel and cause an explosion	Fixing an earth strap between the aeroplane and ground will transfer electrons to the ground preventing the aeroplane becoming charged

3. Any suitable example such as electrostatic scrubbers in power station chimneys, vacuum cleaners, inkjets and photocopiers.

4. They can both attract or repel objects.

Activity P2d.3 Consolidation and summary

Learning objectives

- To review the learning points of the topic
- To test understanding through answering questions

Learning outcomes

- Be familiar with the knowledge and understanding summarised in the End of topic checklist
- Be able to apply this knowledge and understanding by answering End of topic questions

Resources

Student Book pages 126–128

Approach

Introduce the activity

Ask students to work with a partner to make a list of key words from this topic. They could then work together to produce a spider diagram showing how the different concepts are linked. They could compare their list with the list of key words given on page 126 of the Student Book. Discuss the checklist given on page 126 and use questioning to see how much of the content students are comfortable with.

Develop the activity

Ask students to work individually to work through the End of topic questions on pages 127–128 of the Student Book without reference to the text. As they work, walk around the classroom observing their answers and questioning them as necessary to find out which questions are causing difficulties.

Bring the activity together

After a set period, ask the students to stop working and discuss any areas of difficulty you observed as you walked round the class. Students should complete any unanswered questions for homework, but you should stress that they should attempt the questions without reference to the text to see how much content they have retained.

Answers

End of topic questions mark scheme

Question	Correct answer	Marks
1 a)	electrons are rubbed off the surface of the plastic onto the cloth static electricity is produced by removing electrons from one insulator to another. An excess of electrons leads to a negative charge	1 1
1 b)	the positively charged rod induces a negative charge on the surface of the paper near to the rod opposite charges then attract	1 1
2 a)	the passenger is charged by friction as her clothes rub against the seat covers the car is also charged by friction with the road and the air touching metal allows the charge to flow to earth	1 1
2 b)	touching a door handle after walking on a synthetic carpet removing clothing	1 1
3	lightning fuelling aircraft. Static electricity can cause explosions in any situations when there are fuels in the gaseous form	1 2
4	charge transfers from hair to comb static electricity crackles charge transferred from comb to paper enables attraction of paper	2 1 2 1
5	nylon positive and rubber negative will attract and so machine will not run smoothly nylon and nylon will repel so machine will run smoothly	1 1 1 1
6	plastic comb/hair duster and polythene polyester clothes and hair leather and rubber and duster/glass rods all collect charge which can be transferred others do not collect charge	1 1 1 1 1 1 1 1

Question	**Correct answer**	**Marks**
7 a)	Jim: anti-static wrist strap	1
	Rashida: anti-static clothing – trousers jackets made of materials that don't build up charge easily when rubbed	1
	boots' anti-static properties	1
	aircraft grounding strap	1
7 b)	links to ground	1
	discharge static safely	1
7 c)	any suggestion where there is a build up of charge	1
	could be hazard to person or environment	1
7 d)	spark igniting fuel or fumes	1
7 e)	any material that does not build up charge – cotton	1
	nylon and leather	1
7 f)	conductive	1
	charge can flow through them	1
7 g)	less humid	1
	charge can't dissipate so easily	1
	Total:	44

Section 2: Answers to exam-style questions

Question	Correct answer	Marks
2 a)	iron has resistance energy transfer (by collisions)	1 1
2 b) i)	$P = V \times I$	1
2 b) ii)	$I = 1100 \div 230$ 4.8 (A)	1 1
2 c)	if current gets too high fuse melts breaking the circuit	1 1 1
2 d) i)	iron has metal case risk of shock	1 1
2 d) ii)	if case becomes live provides low resistance path so large current/fuse melts	1 1 1
2 e)	a.c. – current changes direction d.c. – electron flow always in one direction	1 1
3 a)	measure current and voltage for different settings of variable resistor	1 1
3 b) i)	line through origin straight initially becoming steeper	1 1 1
3 b) ii)	as current increases wire becomes hotter so resistance increases so gradient increases	1 1 1
3 c)	first part of graph/straight line heating effect not enough to change resistance	1 1 1
3 d) i)	$V = I \times R$	1
3 d) ii)	$3 \div 15$ 0.2 (A)	1 1
3 d) iii)	$Q = I \times t$	1
3 d) iv)	$0.2 \times (10 \times 60)$ 120 C	1 1 1

Question	Correct answer	Marks
4 a)	electrons transferred	1
	gain = negative/loss = positive	1
4 b)	conductor	1
	allows electrons to leave/arrive	1
4 c)	method to allow one rod to move freely (e.g. hang from thread)	1
	bring second rod close	1
	move together = attraction/move apart = repulsion	1
	unlike charges attract/like charges repel	1
4 d)	suitable example	1
	hazard	1
	risk reduction	1
	Total:	44

Section 3. Waves

Contents

Overview of the section

This section considers general properties of waves. The difference between longitudinal and transverse waves is explained, before students learn the various terms associated with waves. The wave equation is introduced and used, as is the relationship between frequency and time period. This content is the basis for the rest of the section.

The second topic considers the electromagnetic spectrum. Students learn the various components of the spectrum in terms of decreasing wavelength and increasing frequency. The topic considers some of the uses of the various radiations, including radio waves (broadcasting and communication), visible light, and X-rays and gamma rays and their applications in medicine.

The final topic considers light and sound. It considers reflection and refraction. Total internal reflection and its use in fibre optic cables is discussed. The topic ends by considering sound. Students learn the frequency range for human hearing, learn about an experiment to find the speed of sound in air, and how an oscilloscope can be used with a microphone to display a sound wave.

Starting points

The Student Book provides a 'section opener', a double-page spread which sets the scene for students (see pages 134–135). It is structured in this way:

- an overview providing details of the areas of study
- five questions, the first three relating to aspects of physics that should be reasonably familiar to most students, and the final two relating directly to some of the topics to be studied
- a list of section contents.

The questions provide a structure for introducing the section.

1. Explain why a red object looks red.

2. Describe the pitch and loudness of the sound you hear if the sound wave has a large amplitude and the frequency is low.

3. Explain the meaning of the words translucent, transparent and opaque.

4. How does light travel through space?

5. Draw a diagram to show how light is reflected in a plane mirror.

The section opener has two main purposes:

- to acknowledge the student's prior learning and to value it
- to provide a benchmark against which future learning can be compared.

The five questions or 'starting points' in the Student Book can be used in a number of different ways to introduce the section:

- You could ask students to consider the questions as an introductory homework task.

- You could put students into groups to share their own ideas and understanding and then report back to the whole class.
- Students could be given access to the Internet, preferably with a tight timescale, to find out the information required.

You could then use a spider chart or other form of wall chart to summarise everybody's ideas. Recording these initial ideas in this way allows you to retain them for reference as the individual topics are developed. In this way, your students' progress in learning can be readily acknowledged.

P3b Properties of waves

Introduction

This topic introduces basic properties of waves and the wave equation.

Links to other topics

Section	Essential background knowledge	Useful links
3 Waves		P3c The electromagnetic spectrum P3d Light and sound

Topic overview

P3b.1	**Properties of waves** This activity lists examples of wave motion. It defines the terms – amplitude, wavelength, frequency and explains the use of these terms to describe waves. Finally, there is a description of transverse and longitudinal wave motion.
P3b.2	**The wave equation** This activity explains the use of the wave equation to link wavelength, frequency and wave speed.
P3b.3	**Reflection and refraction** In this activity students will explore the use of water waves to show reflection at a plane surface and refraction due to change of speed.
P3b.4	**Consolidation and summary** This activity provides an opportunity for a quick recap on the ideas encountered in the topic as well as time for the students to answer the End of topic questions in the Student Book.

Activity P3b.1 Properties of waves

Specification reference: P3b.3.2, P3b.3.3, P3b.3.4

Learning objectives

- Explain the difference between transverse and longitudinal waves
- Define the terms amplitude, wavefront, frequency, wavelength and period of a wave
- Understand that waves transfer energy and information without transferring matter

Learning outcomes

- State examples of wave motion, including light, sound, water and mechanical waves, distinguishing them as longitudinal or transverse
- Use the terms amplitude, wavefront, wavelength, frequency, period and speed to describe waves, including correctly labelling diagrams

Resources

Student Book pages 137–139

Worksheet P3b.1a Making and seeing waves

Worksheet P3b.1b Waves sheet

Resources for class practical and demonstration (see Technician's notes, following)

Approach

Introduce the activity

Discuss examples the students can think of where the idea of 'waves' appears in everyday life. This could include 'scientific' examples they already know, such as light, but draw out a wider range – such as 'Mexican wave' or 'brainwave' – take all answers without comment at this stage.

Develop the activity

Demonstrate wave formation or pulses in the lab using a 'slinky' spring or a rope as shown on page 138 of the Student Book. Point out the key differences between transverse and longitudinal waves and give examples of each type.

Introduce key terms – wavelength, speed, frequency, time period and amplitude – as methods to describe particular wave pulses.

Students can then carry out the practical on Worksheet P3b.1a to see these for themselves. The method for the first part of the practical is as follows. Place the spring on the table. Two students stretch it to a length of about 1 m. Give one end of the spring a quick flick to one side. The pulse will move down the spring and reflect off the hand of the further person. Place the bottle on the table next to the spring about halfway along it. Send pulses down the spring to knock the bottle over.

Shake the spring from side to side to make a transverse wave in the spring. Change the size and speed of the sideways oscillation. Shake the spring towards and away from you to make a longitudinal wave in the spring. Change the size and speed of the backwards and forwards oscillation.

For the second part, Worksheet P3b.1b, the method is as follows:

Cut out pictures of the pulse moving from left to right along the rope and four extra blank pieces of card the same size.

Stack in the correct order with the blanks at the back.

Hold firmly with the left hand and flick with the right so that the pulse can be seen to move along the rope.

Rearrange so the energy flows from right to left when the cards are flicked.

Introduce a 'check list' to check whether a particular example should be counted as a wave. As a guide, use these five properties:

- waves are caused by a vibration or oscillation
- waves transfer energy (and information) without transferring matter
- waves reflect
- waves refract
- waves interfere/superpose.

Introduce the idea of a wavefront by demonstrating water waves in a ripple tank.

Bring the activity together

Get the students to demonstrate a Mexican wave as seen at sporting events. Discuss the nature of the wave (longitudinal or transverse) and the energy transfer (kinetic).

Technician's notes

Be sure to check the latest safety notes on these resources before proceeding.

The following resources are needed for the class demonstration, P3b.1:

slinky spring or rope
ripple tank

The ripple tank will need to be set up according to the manufacturer's instructions.

The following resources are needed for the class practical P3b.1a, per group:

slinky spring
small empty plastic bottle
scissors
P3b.1b Waves sheet, printed on thin card

Answers

Page 139

1. a) The waves travel by vibrations in the direction of travel of the wave.
   b) The waves travel by vibrations at a right angle to the direction of travel of the wave.
2. a) The distance between consecutive peaks or troughs of the wave.
   b) The number of vibrations per second or number of peaks or troughs that pass a point each second.
   c) The size of the vibrations.
3. 15 m.
4. Energy and information.

P3b.1a Making and seeing waves

You are going to demonstrate transverse and longitudinal waves.

Apparatus

slinky spring

empty plastic bottle

scissors

P3b.1b Waves sheet, printed on card

Method

Making waves

1. Work with a partner. Place the spring on the table and stretch it to a length of about 1 m between you and your partner.
2. Give your end of the spring a quick flick to one side. Admire the pulse as it moves down the spring and reflects off your partner's hand.
3. Place the bottle on the table next to the spring about halfway along it. Send pulses down the spring to knock the bottle over.
4. Shake the spring from side to side to make a transverse wave in the spring.
5. Note what happens as you change the size and speed of the sideways oscillation of your hand.
6. Shake the spring towards and away from you to make a longitudinal wave in the spring.
7. Note what happens as you change the size and speed of the backwards and forwards oscillation of your hand.

Seeing waves

1. Cut out the pictures of the pulse moving from left to right along the rope.
2. Cut out four extra blank pieces of card the same size as the others.
3. Stack them together in the right order, with the blank ones at the back.
4. Hold firmly with your left hand and flick through with your right hand so that you can see the pulse move along the rope.
5. Rearrange the order of the pictures so that the energy flows from right to left as you flick through.

Analyse and interpret data

1. What do you need to do to the spring before pulses travel along it?
2. What does shaking the spring side to side do?
3. What does shaking the spring backwards and forwards do?

P3b.1b Waves sheet

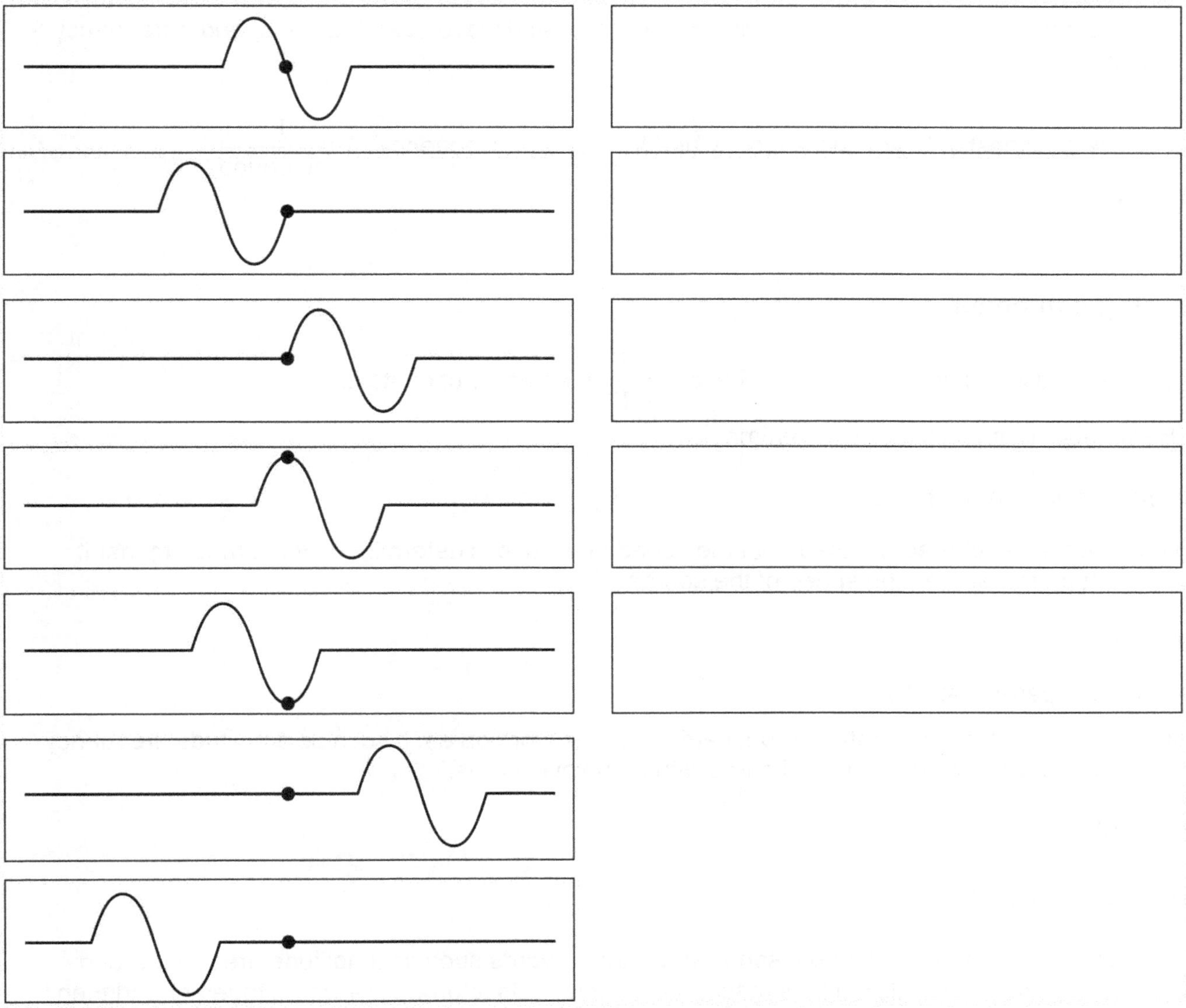

Activity P3b.2 The wave equation

Specification reference: P3b.3.5, P3b.3.6, P3b.3.7

Learning objectives

- Use the wave equation to link wavelength, frequency and wave speed: wave speed = frequency × wavelength
  $v = f \times \lambda$
- Use the relationship between frequency and time period: $\text{frequency} = \frac{1}{\text{time period}}$

  $f = \frac{1}{T}$

Learning outcomes

- Be able to use the equations $v = f \times \lambda$ and $f = \frac{1}{T}$ in a variety of contexts
- Remember that speed = distance/time

Common misconceptions

Some students do not readily accept that the speed of a wave is determined by the medium that it travels through rather than the speed of the source.

Resources

Student Book pages 140–141

Resources for class demonstration: rope, cards with one word on each such as amplitude, frequency, mass and cards with wave words and cards with non-wave words.

Approach

Introduce the activity

Give the students a number of cards, some with 'waves' words such as amplitude, frequency, and some with 'non-waves' words such as mass. Ask the students to group the words together and then extend by calling for the correct definitions to go with the words.

Develop the activity

Introduce the idea of 'wave speed' by sending a pulse along a rope. If you ask students to do this, ask them to try to send the pulse at different speeds – as long as the tension remains constant they should find this very difficult to do. Use a ripple tank if available to illustrate further and extend to derive the connection between wave speed, frequency and wavelength.

Reinforce by asking the students to complete a number of examples. The worked examples on page 140 of the Student Book are good starting points for this. Ask students to tackle the problems without the given solutions initially then, after a set period of time, compare students' solutions with the given solutions and discuss any differences.

Extend by using numbers relating to the electromagnetic spectrum. The frequencies for these waves can be very large (and the wavelengths very small) so students may need assistance to find efficient methods of handling them.

Students can then tackle the questions on page 140 of the Student Book in pairs.

Move on to discuss the relationship between frequency and time period. The worked example on page 141 of the Student Book can be approached as described above before students tackle the questions which follow in pairs.

Bring the activity together

Remind students that speed can also be calculated from

speed = distance travelled ÷ time taken

and demonstrate that in some situations information may be more easily obtained using this method.

Technician's notes

Be sure to check the latest safety notes on these resources before proceeding.

The following resources are needed for the class practical P3b.2, per group:

rope
cards with one word on each such as amplitude, frequency, mass and cards with wave words and cards with non-wave words
ripple tank if available

Answers

Page 140

1. 1.2 m.
2. 3×10^8 m/s.
3. 1500×10^6 Hz.

Page 141

1. 0.002 27 s.
2. 8.33×10^{-7} s.
3. 4×10^{-8} s.

Activity P3b.3 Reflection and refraction

Specification reference: P3b.3.8, P3b.3.9

Learning objectives

- Explain the change in observed frequency and wavelength when the wave source is moving relative to the observer (the Doppler effect)
- Explain reflection and refraction in waves

Learning outcomes

- Be able to relate the change in frequency as a result of the Doppler effect to movement towards and away from the observer
- Describe the use of water waves to show the behaviour of wavefronts, reflection at a plane surface and refraction due to change of speed

Resources

Student Book pages 141–145

Resources for a demonstration (see Technician's notes, following)

Approach

Introduce the activity

Ask students to work with a partner to write down what they understand by the words wavefront, reflection and refraction. Take feedback and share ideas with the class – e.g. post ideas on a poster or a whiteboard.

Develop the activity

Explain that, in this lesson, you are going to demonstrate the behaviour of wavefronts, reflection and refraction using water waves and that these topics will also be studied more thoroughly in a later topic.

Use a ripple tank to demonstrate the following:

- To show reflection, put a plane surface in the ripple tank some distance from the paddle.
- To show refraction, put a clear glass block in the water to change the depth of the water in a given region.

Begin by demonstrating reflection at a plane surface. Discuss how reflection can be explained using wavefronts.

Now demonstrate refraction by using a clear glass block to change the depth of the water in a given region. Discuss the change of speed of the wave using page 142 of the Student Book to help you. Again, discuss how refraction can be explained using wavefronts.

The idea of the Doppler effect can be introduced here. Ask the students 'What would happen to the waves and wavefronts if a source was moving?' Demonstrating this in a ripple tank is difficult. However, a demonstration is possible with a Doppler ball. Alternatively, a constant frequency sound source can be used, for example, a small buzzer with its own battery supply placed inside a small container. The sound source can be turned on and held stationary. It can then be moved towards and away from the students and they can try to recognise the changes in pitch. They can then be asked to try to explain the changes in frequency using the ideas of wavefronts and wavelengths. They can usefully and quickly research this idea by web-searching 'Doppler demonstration'. This will give several places where graphics are used to explain the changes in wavelength and frequencies due to the Doppler effect. If students have ever watched an F1 motor race, then the sound of a racing car going past a fixed microphone demonstrates the Doppler effect very well.

Students should tackle the questions on page 143 of the Student Book and the Developing investigative skills feature on page 144 in pairs.

Discuss the Science in context feature on page 145 of the Student Book.

Bring the activity together

Ask students to make a summary of the main points of the activity in a form of their choice.

Technician's notes

Be sure to check the latest safety notes on these resources before proceeding.

The following resources are needed for the class demonstration, P3b.3:

ripple tank filled with water – set up according to the manufacturer's instructions	clear glass block
plane surface	Doppler ball if available, or a small buzzer in a container

Answers

Page 143

1. An echo is a single reflection from a hard surface.
2. $v = f\lambda$, f constant, so if v reduced so is λ.
3. It depends on the change in speed.
4. a) D
   b) Natural frequency of sound from the train is 410 Hz (frequency when passing the observer).

   Frequency is increased when train approaches.

   Frequency is reduced when train moves away. Steady speed means the increase in frequency must be the same as the decrease in frequency (i.e. 30 Hz is gained on approach and 30 Hz is reduced when moving away).

Developing investigative skills, page 144

1. The students are measuring short time intervals (as the speed of sound is quite fast). Their measurements also depend on the first student tapping out a rhythm exactly in time with the echoes. Reaction time is looked at in question 6. All these elements suggest that anomalous results will be quite easy to generate, so repeat measurements, and plenty of them, would be wise in this case.

2. If the wall is 100 m away then the echo is obviously heard when the sound has travelled 200 m. This still takes less than one second and the timing will be quite difficult to do accurately by hand. However, if the distance is much further, then the energy from the sound wave will have dispersed over a wider area and the echo will be more difficult to hear. 100 m seems a reasonable compromise.

3. Speed of sound = distance travelled ÷ time taken = (4 × 200 m)/2.3 s = 350 m/s (to 2 s.f.).

4. To generate data for a graph, the students could measure the times between different numbers of echoes. For example:

Number of echoes	Distance (m) (number of echoes × 200)	Time interval (s)

This allows a graph to be plotted of distance on the *y*-axis against time on the *x*-axis. The speed of sound is then the gradient of the graph.

5. As mentioned in question 1, there are a number of factors which could cause an individual measurement to be inaccurate. Plotting a graph would allow such measurements to be picked up more easily and the best fit line would allow such anomalies to be smoothed out.

6. Reaction times could affect the tapping out of the rhythm on the metal bar, since the student has to hear the echo to react to it. Also, the student timing has to react at the start of the measurement and at the end. All these reactions add uncertainty to the measurements.

7. A sound sensor connected to a data logger or a computer could reduce the effect of at least some of the reaction times.

Activity P3b.4 Consolidation and summary

Learning objectives

- To review the learning points of the topic
- To test understanding through answering questions

Learning outcomes

- Be familiar with the knowledge and understanding summarised in the End of topic checklist
- Be able to apply this knowledge and understanding by answering End of topic questions

Resources

Student Book pages 146–147

Approach

Introduce the activity

Ask students to work with a partner to make a list of key words from this topic. They could then work together to produce a spider diagram showing how the different concepts are linked. They could compare their list with the list of key words given on page 146 of the Student Book. Discuss the checklist given on page 146 and use questioning to see how much of the content students are comfortable with.

Develop the activity

Ask students to work individually to work through the End of topic questions on page 147 of the Student Book without reference to the text. As they work, walk around the classroom observing their answers and questioning them as necessary to find out which questions are causing difficulties.

Bring the activity together

After a set period, ask the students to stop working and discuss any areas of difficulty you observed as you walked round the class. Students should complete any unanswered questions for homework, but you should stress that they should attempt the questions without reference to the text to see how much content they have retained.

Answers

End of topic questions mark scheme

<table>
<tr><th>Question</th><th>Correct answer</th><th>Marks</th></tr>
<tr><td>1 a)</td><td>the diagram should show several sine waves:
the crest is the top of the wave
the wavelength is the length of the repeating pattern
the amplitude is half the distance between the crest and the trough</td><td>
1
1
1</td></tr>
<tr><td>1 b)</td><td>512 waves per second</td><td>1</td></tr>
<tr><td>2 a)</td><td>0.33
m</td><td>1
1</td></tr>
<tr><td>2 b)</td><td>1.11 × 10–9
s</td><td>1
1</td></tr>
<tr><td>3</td><td>0.75
m</td><td>1
1</td></tr>
<tr><td>4</td><td>3×10^9
Hz</td><td>1
1</td></tr>
<tr><td>5</td><td>80
m/s</td><td>1
1</td></tr>
<tr><td>6</td><td>completed table as shown
<table>
<tr><th>Speed (m/s)</th><th>Frequency (Hz)</th><th>Wavelength (m)</th><th>Period (s)</th></tr>
<tr><td>20</td><td>10</td><td>2</td><td>0.1</td></tr>
<tr><td>20</td><td>5</td><td>4</td><td>0.2</td></tr>
<tr><td>30</td><td>5</td><td>6</td><td>0.2</td></tr>
<tr><td>340</td><td>10</td><td>34</td><td>0.1</td></tr>
<tr><td>160</td><td>8</td><td>20</td><td>0.125</td></tr>
<tr><td>3000</td><td>250</td><td>12</td><td>0.004</td></tr>
</table></td><td>12</td></tr>
</table>

Question	Correct answer	Marks
7 a)	the water gets shallower 200 m ahead	1 1 1
7 b)	the current is going right to left the waves make the piece of wood bob up and down but they do not move it along	1 1 1
8	diagram to show wavefronts moving out from the plane wavefronts closer together when approaching wavefronts further apart when leaving	1 1 1
	Total:	35

P3c The electromagnetic spectrum

Introduction

This topic introduces students to the electromagnetic spectrum.

Links to other topics

Section	Essential background knowledge	Useful links
3 Waves	P3b Properties of waves	P3d Light and sound
6 Magnetism and electromagnetism		P6c Electromagnetism
7 Radioactivity, fission and fusion		P7b Radioactivity

Topic overview

P3c.1	**Electromagnetic spectrum** This activity teaches students to state the regions of the electromagnetic spectrum in order of increasing frequency (decreasing wavelength) and vice versa. It describes the sources of the different regions of the spectrum and then describes in detail the uses of some regions of the spectrum.
P3c.2	**Uses and hazards of electromagnetic radiation** This activity describes in detail the uses of each region of the spectrum from the last activity. It then describes hazards relating to the spectrum and some measures to reduce risks to humans.
P3c.3	**Consolidation and summary** This activity provides an opportunity for a quick recap on the ideas encountered in the topic as well as time for the students to answer the End of topic questions in the Student Book.

Activity P3c.1 Electromagnetic spectrum

Specification reference: P3b.3.10, P3b.3.11

Learning objectives

- Know that light is part of the electromagnetic spectrum that includes radio, microwave, infrared, visible, ultraviolet, X-ray and gamma ray radiations
- Know that all these waves travel at the same speed in free space
- Know the regions of the electromagnetic spectrum in order of increasing frequency (decreasing wavelength), including the colours of the visible spectrum

Learning outcomes

- Be able to list the regions of the electromagnetic spectrum in order, from radio waves to gamma radiation

Common misconceptions

The disturbances associated with electromagnetic waves (electric and magnetic fields) are invisible, so students may have difficulty in grasping exactly what the amplitude of an electromagnetic wave is.

Some students think that we listen to radio waves.

Resources

Student Book pages 148–154

Worksheet P3c.1a The colours of light

Resources for the class practical (see Technician's notes, following)

Approach

Introduce the activity

Line up the students so that they perform Mexican waves of different wavelengths. Revisit the ideas of frequency, wavelength and wave speed.

Develop the activity

Discuss 'A family of waves' on pages 148–149 of the Student Book. Students will be most familiar with the visible part of the spectrum.

Students then carry out the practical on Worksheet P3c.1a. The practical gives them the opportunity to obtain a spectrum of visible light. Note that students should take care with the glass prism.

Students need to know the order of the spectrum in terms of increasing wavelength and increasing frequency/energy. It is important that students realise that all electromagnetic waves travel at the same speed in a vacuum (3×10^8 m/s). Students should tackle the questions on page 150 of the Student Book in pairs.

Discuss the Science in context feature on pages 150–151. Talk about uses of lasers.

Bring the activity together

Ask students to work with a partner to write down two things that they think went well during this activity and one thing that they would like to have done differently. Take feedback and discuss any areas of difficulty.

Technician's notes

Be sure to check the latest safety notes on these resources before proceeding.

The following resources are needed for the class practical P3c.1a, per group:

ray box, power supply and leads
cylindrical lens
ray box comb
wooden block
sticky putty
white card
mirror
glass prism

This practical works best in a darkened room.

SAFETY INFORMATION

Take care with the glass prism and the ray box lamp, which may get hot.

Ensure laboratory is clear of any objects students may trip on in the dark.

Answers

Page 150

1. A prism splits light by using the fact that different wavelengths of light slow down inside the glass by different amounts so they are refracted through different angles.

2. 3×10^8 m/s.

3. X-rays.

4. Gamma, UV, visible, infrared, microwaves.

P3c.1a The colours of light

In this practical, you will split light through a prism to view a spectrum.

Apparatus

mirror
wood block
sticky putty (e.g. *Blu-tack*)
ray box and comb
power supply and leads
glass prism and cylindrical lens white card

> **SAFETY INFORMATION**
> *Take care handling the glass prism and lenses.*
> *Also the ray box may get hot.*
> *Do not move around the laboratory in the dark.*

Method

1. Fasten the card onto the block with sticky putty to make a screen.
2. Place the ray box about a metre from the screen. Use the cylindrical lens to focus the lamp filament onto the screen.

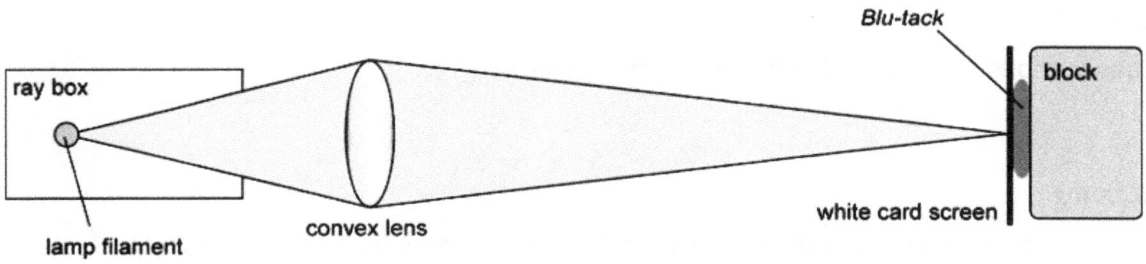

3. Place the prism on the bench as shown below. Move the screen round to obtain a clear spectrum of rainbow colours.

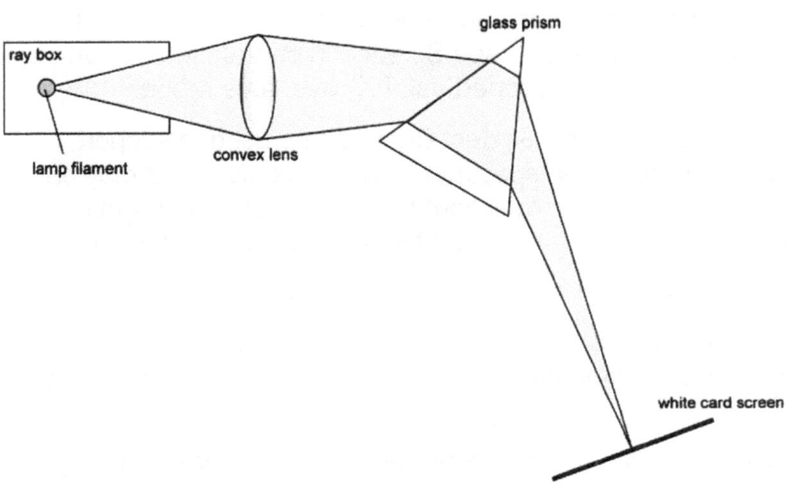

Activity P3c.2 Uses and hazards of electromagnetic radiation

Specification reference: P3b.3.12, P3b.3.13

Learning objectives

- Explain in detail uses of electromagnetic radiation
- Explain hazards relating to excessive exposure to electromagnetic waves and describe some measures to reduce risks to humans

Learning outcomes

- Describe at least one use of each region of the electromagnetic spectrum
- Recall possible harmful effects on humans from relevant regions of the electromagnetic spectrum
- Explain methods of reducing the risks to humans from waves of the electromagnetic spectrum

Common misconceptions

Many students may have received X-rays in the past – usually either for suspected fractures and broken bones, or for dental reasons. Due to their widespread use, many students automatically assume that the procedure is inherently safe. While we are not intending to contradict this assumption, students should be made aware that any exposure to ionising radiation is considered harmful. The risk to the individual is almost negligible. However, when considered against the size of a population, ethical issues need to be considered.

Resources

Student Book pages 150–154

Resources for class demonstration (see Technician's notes, following)

Approach

Introduce the activity

Ask students to write down the seven groups of waves in the electromagnetic spectrum, from the lowest frequency through to the highest frequency waves.

Develop the activity

Discuss the position of X-rays within the electromagnetic spectrum. Explain to students about the properties of gamma rays, which are ionising radiation which can cause cancer in human cells, and ultra violet radiation, exposure to which is linked with skin cancer (see Science in context feature on page 152 of the Student Book). Ask students to deduce some properties of X-rays, as a result.

The diagram of the electromagnetic spectrum on page 149 of the Student Book can be used to support students who are unable to recall the position of waves within the spectrum. To avoid confusion, it may be helpful to remind students that X-rays (and other electromagnetic radiation) travel at the speed of light in a vacuum and just slightly slower in air, which is why the same value is used for calculations at this level.

Discuss how an X-ray image is formed.

Discuss further with students the potentially harmful effects of ionising radiation. Explain that professionals who work with X-rays wear film badges or dosimeters to monitor the dose of radiation to which they have been exposed.

Move on to discuss microwaves. Students may well have experience of the use of microwaves to cook food but will probably be unaware of precisely how the process works. If possible you could demonstrate cooking a small potato in a microwave oven during the lesson.

The final section of the spectrum to be discussed is radio waves. Point out the various uses of different wavelengths of radio waves as described on page 153 of the Student Book. Students can then tackle the questions on page 153 in pairs.

More able students can tackle the extension feature on page 154 of the Student Book.

Bring the activity together

Draw the table below onto the board for students to copy down. Ask students to work with a partner to complete the table.

Electromagnetic radiation	Uses	Hazards associated with it
Radio waves		
Microwaves		
Infrared		
Visible light		
Ultraviolet		
X-rays		
Gamma rays		

Technician's notes

Be sure to check the latest safety notes on these resources before proceeding.

The following resources are needed for the class demonstration P3c.2:

microwave oven
potato

Do not allow students to eat anything in a laboratory.

Answers

Page 153

1. Gamma rays are produced by radioactive nuclei. X-rays are produced by firing high energy electrons at a metal target.

2. An image of infrared radiation.

3. Water particles in food absorb the energy carried by microwaves. They vibrate more making the food much hotter.

Page 154

Students' own answers.

Activity P3c.3 Consolidation and summary

Learning objectives

- To review the learning points of the topic
- To test understanding through answering questions

Learning outcomes

- Be familiar with the knowledge and understanding summarised in the End of topic checklist
- Be able to apply this knowledge and understanding by answering End of topic questions

Resources

Student Book pages 155–157

Approach

Introduce the activity

Ask students to work with a partner to make a list of key words from this topic. They could then work together to produce a spider diagram showing how the different concepts are linked. They could compare their list with the list of key words given on page 155 of the Student Book. Discuss the checklist given on page 155 and use questioning to see how much of the content students are comfortable with.

Develop the lesson

Ask students to work individually to work through the End of topic questions on pages 156–157 of the Student Book without reference to the text. As they work, walk around the classroom observing their answers and questioning them as necessary to find out which questions are causing difficulties.

Bring the lesson together

After a set period, ask the students to stop working and discuss any areas of difficulty you observed as you walked round the class. Students should complete any unanswered questions for homework, but you should stress that they should attempt the questions without reference to the text to see how much content they have retained.

Answers

End of topic questions mark scheme

Question	Correct answer	Marks
1 a)	light	1
1 b)	ultraviolet	1
1 c)	microwaves	1
1 d)	X-rays	1
1 e)	infrared	1
2 a)	they can pass through soft tissue and kill cancer cells: gamma rays are useful because of their great penetrating power this is also their disadvantage consequently they have to be used extremely carefully	1 1 1
2 b)	they can damage healthy cells and cause cancer because of their very high energy	1 1 1
2 c)	they have different frequencies and wavelengths	1
2 d)	300 000 000 m/s	1
3 a)	radio, microwaves, infrared, visible, ultraviolet, X-rays, gamma	4
3 b)	the order would be reversed	1
4	heat energy heats air surrounding heater causing convection current which then warms air around hands	1 1 1

Question	**Correct answer**	**Marks**
5	completed table as shown	13

Wave type	**Source of wave**	**Use**	**Property**
radio	radio transmitter and aerial	**communication**	**low-energy waves**
X-rays	X-ray tubes	security at airports	**penetrate tissue**
visible light	lamps, Sun and flames	**seeing**	can be dispersed into seven colours
ultraviolet	mercury vapour lamps	security markings	**causes some chemicals to fluoresce**
infrared	**all warm objects**	thermal imaging	can cause burns
gamma	radioactive substances	**destroying cancer cells**	very penetrating
sound	**vibrating objects**	hearing	**travels through solids, liquids and gases**
microwaves	magnetron	heating food quickly	**cause water molecules to vibrate**

Question	**Correct answer**	**Marks**
6	microwaves cause water molecules in food to vibrate more than in cold state raising internal energy of food and thus cooking it	1 1 1
7	radio waves carry audio and visual information through distance to receiver where the information is converted back to audio and visual	1 1 1
8	radio waves will be diffracted over the hill and be received at the house better than short wave television waves so radio reception will be better	1 1 1
	Total:	43

P3d Light and sound

Introduction

This topic covers perhaps the most familiar types of waves: light and sound.

Links to other topics

Section	Essential background knowledge	Useful links
3 Waves	P3b Properties of waves P3c The electromagnetic spectrum	

Topic overview

P3d.1	**Reflection of light** This activity describes how light is reflected from shiny surfaces. Students are asked to recall the law of reflection and to apply the law of reflection to explain the formation of an image in a plane mirror.
P3d.2	**Refraction of light** This activity describes what happens when a ray of light is refracted at the boundary between two media. It links the direction changes involved in refraction correctly to the densities of the two media and then discusses Snell's law and the link to refractive index.
P3d.3	**Total internal reflection and critical angle** This activity looks at refraction at a more dense/less dense boundary. It introduces the critical angle and total internal reflection.
P3d.4	**Properties of sound** This activity applies wave ideas to the phenomenon of sound. It examines the range of human hearing and introduces ultrasound as sounds at higher frequencies. It considers simple applications of ultrasound.
P3d.5	**Measuring the speed of sound** This activity describes a method to allow the speed of sound to be measured.
P3d.6	**Consolidation and summary** This activity provides an opportunity for a quick recap on the ideas encountered in the topic as well as time for the students to answer the End of topic questions in the Student Book.

Activity P3d.1 Reflection of light

Specification reference: P3b.3.14, P3b.3.15, P3b.3.16

Learning objectives

- Know that light waves are transverse waves that can be reflected
- Use the law of reflection (the angle of incidence equals the angle of reflection)
- Draw ray diagrams to illustrate reflection

Learning outcomes

- Describe how light is reflected from plane surfaces, including using ray diagrams
- Be able to construct ray diagrams to show how images are formed in plane mirrors

Resources

Student Book pages 158–162

Worksheet P3d.1a Reflection

Worksheet P3d.1b Images from rays

Worksheet P3d.1c Making a periscope

Resources for class practicals (see Technician's notes, following)

Approach

Introduce the activity

Ask students to work with a partner to discuss where they see reflection in everyday life. Take feedback.

Develop the activity

Discuss the content of Student Book page 159, including the Science in context feature. Students could carry out the practical on Worksheet P3d.1a or you could do this as a demonstration as long as students are fully involved if they have not already done this. The experiment works best in a darkened room.

Set up the ray box so that it sends a single beam of light across the white paper. Fasten the mirror to the wooden block with sticky putty and place on the paper so that it reflects the incident beam of light. Mark the beams on paper, draw in the normals and measure the angles of incidence and reflection. Students often find measuring angles with a protractor difficult. Encourage them to start off with 0 degrees lined up on the normal.

Now consider 'Reflection of light and ray diagrams' on page 160 of the Student Book. Take time to make sure that students fully understand the concepts introduced. At this point students can carry out the practical on Worksheet P3d.1b.

Now consider the use of a periscope. Students should carry out the practical on Worksheet P3d.1c. Discuss the Science in context feature on page 161–162 of the Student Book. Finally, students can tackle the questions on page 162 of the Student Book.

Bring the activity together

Ask students to make a summary of the main points of the activity in a form of their choice. They could work in pairs to do this.

Technician's notes

Be sure to check the latest safety notes on these resources before proceeding.

The following resources are needed for the class practical P3d.1a, per group:

ray box, power supply and leads
cylindrical lens
ray box comb
wooden block
sticky putty
plastic coated mirror
protractor
white card

This practical works best in a darkened room.

Take care. Ray boxes become hot after prolonged use.

The following resources are needed for the class practical P3d.1b, per group:

squared paper	plastic coated mirror
protractor	power supply
50 cm ruler	ray box, comb and cylindrical lens
sticky putty	two 4 mm leads

This practical works best in a darkened room.

Take care. Ray boxes become hot after prolonged use.

The following resources are needed for the class practical P3d.1c, per group:

pair of plastic mirrors
50 cm ruler
sticky putty

Answers

Page 162

1. 48°.

2. The 'normal' is the imaginary line at 90° to a mirror.

P3d.1a Reflection

Apparatus

plastic coated mirror
wood block
sticky putty
white paper
ray box, lens and comb
leads and power supply
protractor
white card

SAFETY INFORMATION
Take care handling the glass prism and lenses.
Also the ray box may get hot.
Do not move around the laboratory in the dark.

Method

Reflection

1. Set up the ray box so that it sends a single ray of light across the white paper.

2. Fasten the mirror to the wooden block with sticky putty and place on the paper so that it reflects the incident beam of light.

3. Carefully draw a pencil line along the front of the mirror. Use a pencil to place crosses under the incident and reflected beams of light.

4. Remove the paper. Draw lines through the crosses to show the incident and reflected beams. Use a protractor to draw in the normal, a line at right angles to the mirror where the reflection happens.

5. Measure the angles *i* and *r*. Are they the same?

6. Repeat steps 1 to 5 for other values of the angle of the mirror to the incident beam.

Analyse and interpret results

1. The three diagrams show waves approaching a reflecting surface.

The incident ray shows the direction that the four wavefronts are moving in. On each diagram, draw
- the normal
- the reflected ray
- the reflected wavefronts

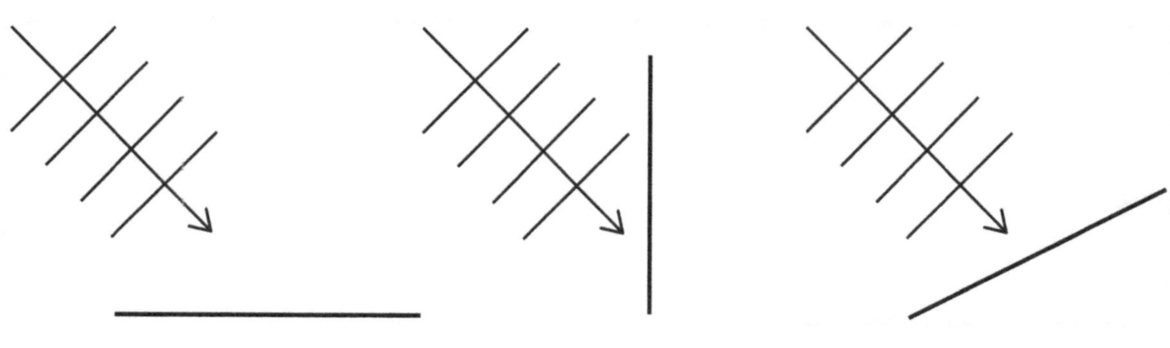

P3d.1b Images from rays

You are going to construct an image in a mirror using rays.

Apparatus

squared paper
ruler
protractor
sticky putty
plastic coated mirror
power supply, pair of leads
ray box, cylindrical lens and comb

SAFETY INFORMATION

Take care. The ray box may get hot.
Do not move around the laboratory in the dark.

Method

1. Draw a 50 mm horizontal line across the centre of the squared paper. Label it mirror.

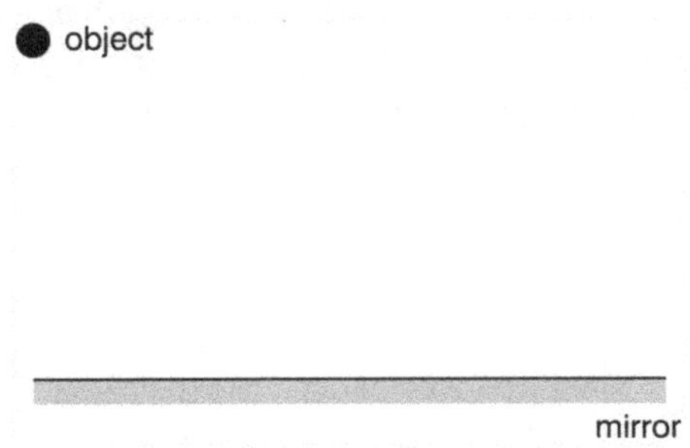

2. Draw a dot 25 mm above the left-hand end of the mirror. Label it object.
3. Draw and label an arrowed line to represent the incident ray from the object to the centre of the mirror.

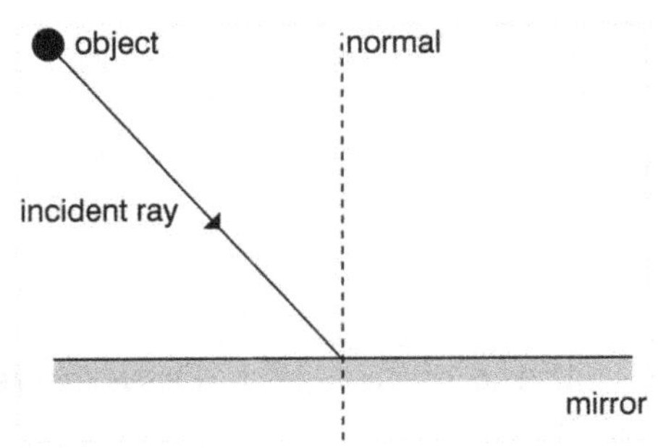

4. Use a protractor to draw and label the normal where the ray hits the mirror.

5. Measure the angle of incidence *i*. Draw a line at the same angle to represent the reflected ray. Continue the ray behind the mirror to show where it appears to come from. The image must lie somewhere on this line.

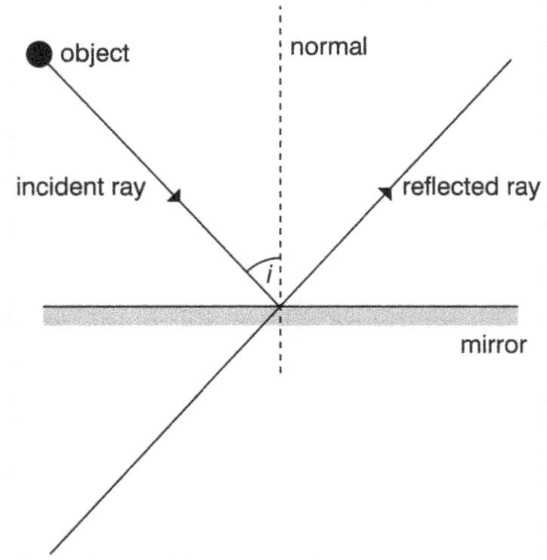

6. Draw a second ray from the object to the mirror. Draw in the normal and use it to draw the reflected ray. Continue this ray behind the mirror. The image must be somewhere on this line as well, so it must be where the two lines cross behind the mirror. Label the image and measure its distance from the mirror. Is it the same distance behind the mirror as the object is in front?

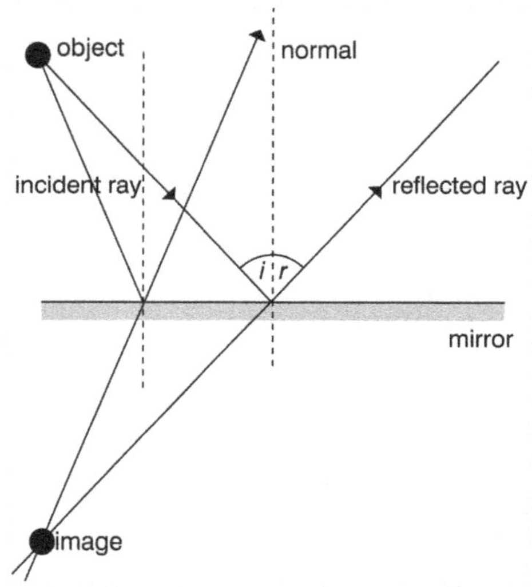

7. Draw a third ray from the object to the far end of the mirror. Draw the reflected ray and continue it back until it crosses the other two behind the mirror.

P3d.1c Making a periscope

You are going to make a periscope using two mirrors.

Apparatus

pair of plastic mirrors

ruler

sticky putty

Method

1. Use sticky putty to fasten a pair of mirrors to a ruler to make a periscope.
2. Adapt your periscope so that you can use it to look behind you, over your head.
3. Hold the two mirrors on the table at right angles with the sticky putty. Look at your image in this arrangement. How is it different from your image in just one mirror?

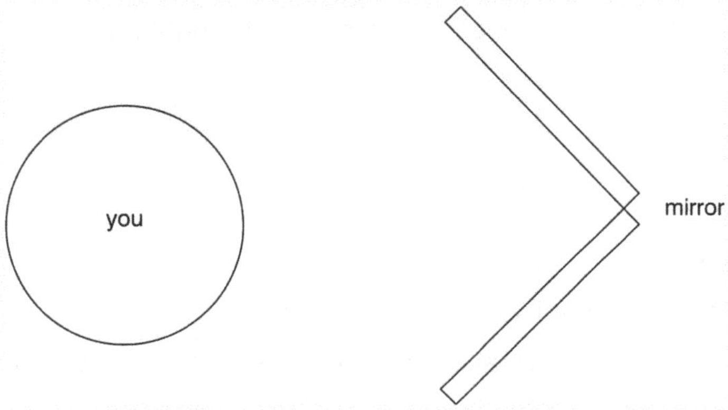

Analyse and interpret results

1. Each diagram shows an object, one or two mirrors and an eye.

Draw a ray of light from the object which reflects off the mirror(s) and enters the eye. Use the ray to indicate the position of the image seen by the eye.

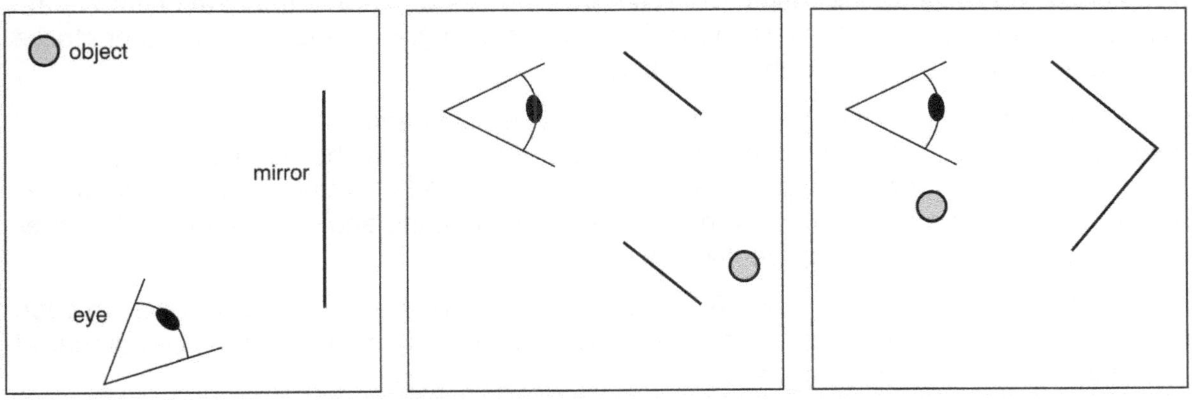

Activity P3d.2 Refraction of light

Specification reference: P3b.3.14, P3b.3.16, P3b.3.17, P3b.3.18, P3b.3.19

Learning objectives

- Know that light waves can be refracted
- Recall and use the relationship between refractive index, angle of incidence and angle of refraction: $n = \frac{\sin i}{\sin r}$
- Determine refractive index of a glass block

Learning outcomes

- Describe an experimental demonstration of the refraction of light
- Describe an experiment to determine the refractive index of glass, using a glass block

Common misconceptions

Students are often surprised by the path light follows through a triangular prism.

Resources

Student Book pages 162–166

Worksheet P3d.2a Refraction

Worksheet P3d.2b Refractive index

Resources for class practicals (see Technician's notes below)

Resources for class demonstration: Glass beaker half filled with water and a long pencil.

Approach

Introduce the activity

Half fill a glass beaker with water and place a long pencil in it. Draw students' attention to the apparent break in the pencil where it enters the water and relate this to refraction.

Develop the activity

Students carry out the practical on Worksheet P3d.2a. Set up the ray box so that it sends a single beam of light across the white paper. Place the rectangular *Perspex*® block in the path of the beam. Mark the beams on paper, draw in the normals and measure the angles. Repeat for the other blocks. The experiments work best in a room with a good blackout.

For more able students, you may wish to carry this out as a demonstration.
Students then carry out the practical on Worksheet P3d.2b. Set up the ray box so that it sends a single beam of light across the white paper. Place the rectangular *Perspex*® block in the path of the beam. Mark the beams on paper, draw in the normals and measure the angles. Repeat for the other blocks. The experiments work best in a room with a good blackout.

Discuss the text Refraction of light on page 162 of the Student Book. The worked examples on page 164 should be tackled without the given solutions initially. Students' solutions can then be compared with the given solution and any differences discussed.

More able students can then tackle the extension feature on page 165 of the Student Book. All students should do the questions on page 165 with a partner.

Students can work in pairs to work through the Developing investigative skills feature on page 166 of the Student Book.

Bring the activity together

Ask students to tell you two things they have learnt in this activity and one thing they found difficult. Discuss areas of difficulty and use the things they have learnt to produce a class summary of the activity.

Technician's notes

Be sure to check the latest safety notes on these resources before proceeding.

The following resources are needed for the class practical P3d.2a, per group:

ray box, power supply and leads
cylindrical lens
ray box comb
protractor
white card
scalene triangle *Perspex®* block
right-angled triangle *Perspex®* block
semicircular *Perspex®* block
glass prism

This practical works best in a darkened room.

Take care. Ray boxes become hot after prolonged use.

The following resources are needed for the class practical P3d.2b, per group:

rectangular glass or *Perspex®* blocks
ray box with thin slit, power supply and leads
plain A4 paper
sharp pencil
protractor

Take care. Ray boxes become hot after prolonged use.

Answers

Developing investigative skills, page 163

1. Shine a narrow beam of light, mark the path of light as it enters the block and as it leaves the block, draw around the transparent block, join the start of the refracted ray to the end of the path of the incident ray, vary incidence angle, measure angles with a protractor from the normal (shown on diagram), repeat for a number of different angles of incidence.

2. independent variable = angle of incidence

dependent variable = angle of refraction

control variable = refractive index of block

3. difficulty in marking the ray (when it is diffuse) so angles not very accurate

Page 165 (top)

1. The light from a submerged object is refracted away from the normal as it leaves the pond, so it approaches your eye at a shallower angle. Your brain is fooled into thinking this shallower angle is the true angle to the object.

2. Being more dense than air, the carbon dioxide in the balloon will slow down the sound wave. This will tend to focus the sound together, like a lens. (A balloon filled with hydrogen will do the opposite.)

3. 30.7°.

4. 36.3°.

5. 1.43.

Page 165 (bottom)

1. Slower.

2. Diamond, because its refractive index is greatest.

3. Air: 2.99×10^8 m/s; water: 2.25×10^8 m/s; glass: 1.97×10^8 m/s; diamond: 1.24×10^8 m/s.

Developing investigative skills, page 166

1. Standard diagram with surface indicated, normal line drawn in and angles of incidence and refraction correctly marked.

2. Draw along the edge of the block while it is in position. With the block removed, find the point where the light entered. At that point, use a protractor to mark in the line at right angles – this is the normal.

3. Independent variable: angle of incidence.
Dependent variable: angle of refraction.

4. Do not touch the hot lamp in the light box or ray box. Take the usual electrical safety precautions when using electrical power supplies.

5. Refractive index = sin i ÷ sin r.

6. Graph should be a straight line through the origin.

7. Refractive index = gradient of the graph.

8. Even though the ray of light is narrow, it still has some width and this can make some of the measurements difficult. In particular, marking the path of the ray can be uncertain (possibly avoided by marking the edge of the ray) and it is difficult to get the normal in exactly the correct position. Correctly drawing in the path inside the block is also difficult to do accurately.

9. A different colour of light would have a different wavelength. This would lead to a different change in speed of the light and so a different angle of refraction (think of how a prism disperses light). However, at the level of precision of this experiment it is unlikely to have a significant effect on the final value calculated.

P3d.2a Refraction

In this practical, you will study refraction.

Apparatus

ray box, lens and comb

leads

power supply

protractor

Perspex® blocks and glass prism

white card

SAFETY INFORMATION

Take care handling the glass prism and lenses.
Also the ray box may get hot.
Do not move around the laboratory in the dark.

Method

1. Set up the ray box so that it sends a single beam of light across the white paper.
2. Place the rectangular *Perspex*® block in the path of the beam.
3. Draw around the block. Place crosses on the beam of light before and after the block.
4. Remove the block. Use the crosses to draw the beam of light as it passed through the block and out the other side. Use a protractor to draw the normals where the light enters and leaves the block.
5. Repeat for these other blocks.

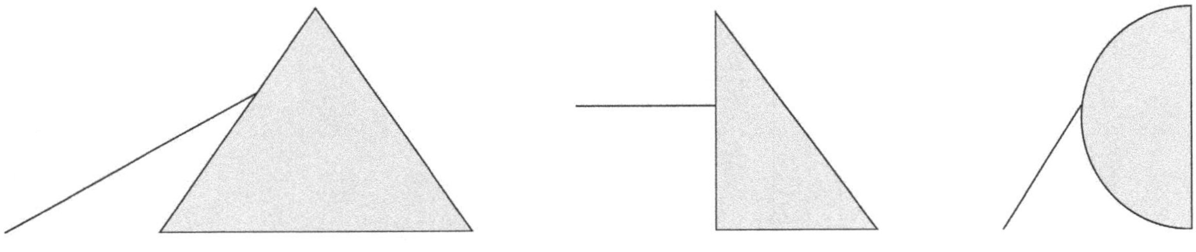

Analyse and interpret data

1. What happens to the wavefronts and the wavelength as light enters a glass block?
2. What happens to the direction of the beam of light as it enters a glass block?
3. What happens to the wavefront and the wavelength as light leaves a glass block?
4. What happens to the direction of the beam of light as it leaves a glass block?

P3d.2b Refractive index

You are going to find the refractive index of a clear block.

Apparatus

power supply and leads
ray box with thin slit
rectangular block
protractor
plain A4 paper
sharp pencil

SAFETY INFORMATION
The ray box may get hot.
Do not move around the laboratory in the dark.

Method

1. Place the rectangular block in the centre of your paper and draw around it.
2. Remove the block and draw a normal line to the block, as shown:

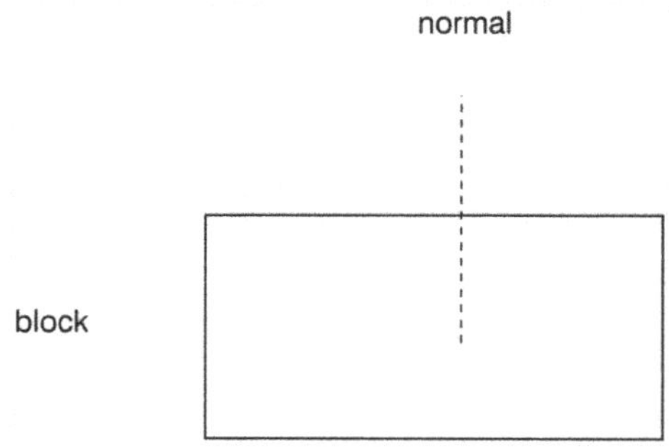

3. Measure an angle of 10° to the normal and draw a line at this angle:

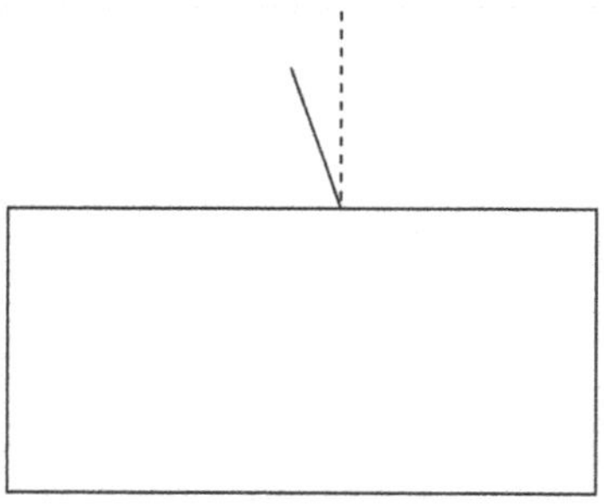

4. Replace the block in position and send a ray of light into the block along the 10° line.

5. Mark using 3 'X' marks where the light emerges from the block and join:

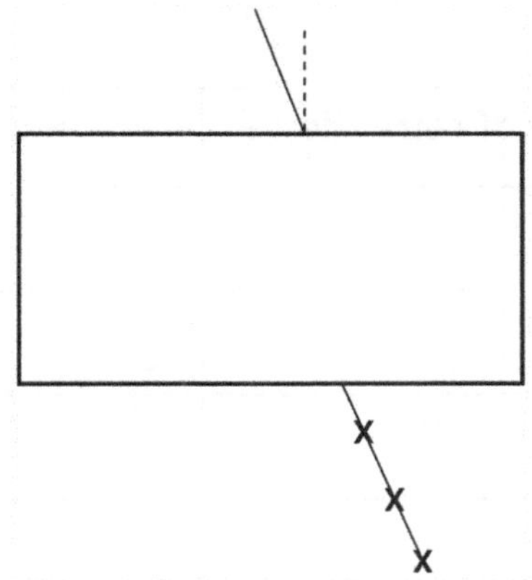

6. Fill in the gap that the light follows inside the block:

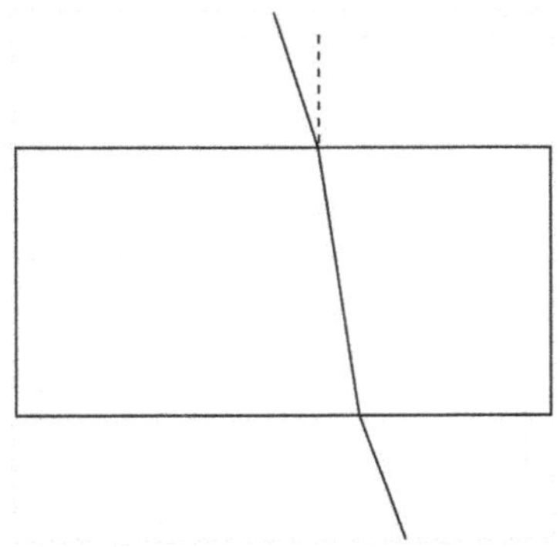

7. Now measure the angle of incidence and angle of refraction:

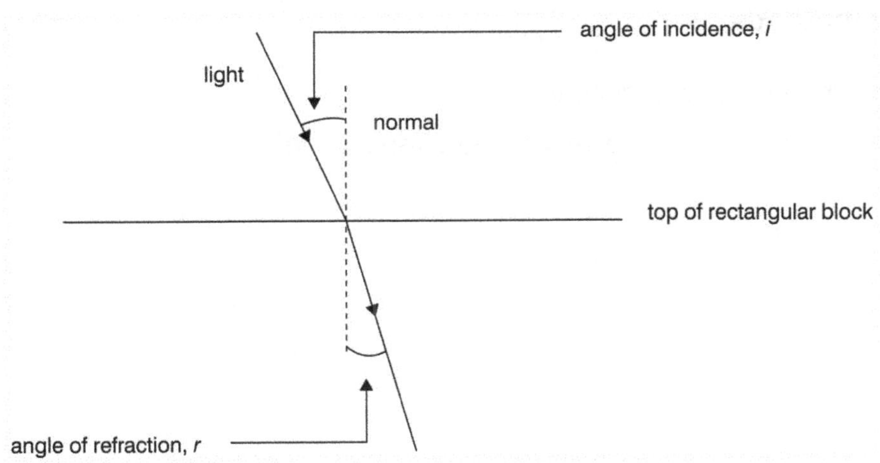

8. Record the values of the angle of incidence (*i*) and the angle of refraction (*r*), in a table like the one below.

9. Repeat the experiment seven more times, increasing the angle of incidence by 10° each time until you reach 80°.

Analyse and interpret data

1. Record your results in a table like this.

Angle of incidence *i*	**Angle of refraction *r***	**sin *i***	**sin *r***
10°			
20°			
30°			
40°			
50°			
60°			
70°			
80°			

2. Plot a graph of sin *i* (*y*-axis) against sin *r* (*x*-axis).
3. What do you notice about the graph?
4. What does the value $\frac{\sin i}{\sin r}$ measure?

Evaluate data and methods

1. List any sources of error in your experiment.
2. How could you make your experimental results more repeatable?

Activity P3d.3 Total internal reflection and critical angle

Specification reference: P3b.3.20, P3b.3.21, P3b.3.22

Learning objectives

- Describe the role of total internal reflection in transmitting information along optical fibres and in prisms
- Introduce total internal reflection and the critical angle
- Know and use the relationship between critical angle and refractive index: $\sin c = \frac{1}{n}$

Learning outcomes

- Give the meaning of critical angle, *c*
- Describe total internal reflection in optical fibres and prisms
- Describe the action of optical fibres particularly in medicine and communications technology

Resources

Student Book pages 167–170

Worksheet P3d.3a Critical angle

Resources for a class practical (see Technician's notes, following)

Resource for class demonstration: An optical fibre lamp.

Approach

Introduce the activity

Ask students to draw diagrams to represent: the reflection of light and the refraction of light through a rectangular glass block.

Develop the activity

Show students an optical fibre lamp. Ask them to attempt to explain how the light works. Remove the fibres from the lamp and shine a bright light (a torch, for example) through the clump of fibres, to demonstrate that it is 'normal' light being transmitted.

Define the term 'critical angle' and then ask students to carry out the practical. Students send the beam from the ray box into a semi-circular block. They should adjust the angle of incidence at the glass-air boundary, to demonstrate the following: refraction; critical angle; total internal reflection.

Challenge students to measure the critical angle accurately (approximately 42°).

Discuss 'Total internal reflection and the critical angle' on pages 167–168 of the Student Book, including the Science in context feature on pages 168–169 on optical fibres and how they are used to transmit signals.

It is useful to extend work on optical fibres to look at the uses of endoscopes. Using the search term 'endoscopy video' will produce a range of interesting practical examples for students to see and discuss.

Discuss the text on page 170 of the Student Book and then ask students to tackle the questions which follow in pairs.

You could also give students the following questions for extra practice:

- A ray of light enters a material at an angle of 36°. It is refracted to an angle of 23°. What is the refractive index? [1.50]
- A ray of light enters a material at an angle of 20°. It is refracted to an angle of 8.2°. What is the refractive index? [2.40]
- A material has a refractive index of 1.42. What is the critical angle? [44.8°]
- The critical angle for flint glass is 38.7°. What is its refractive index? [1.60]
- A ray of light enters Pyrex® glass at an angle of 42.9°. The light is refracted along the Pyrex®-air boundary. What is the refractive index of Pyrex®? [1.47]
- The critical angle of a type of crown glass is 36°. To what angle would light be refracted if it were incident on the glass at an angle of 70°? [33.6°]

Bring the activity together

Ask students to define the term 'critical angle'. Ask students to produce a series of three diagrams which demonstrate the possible paths of a ray of light as it enters a semi-circular glass block. Diagrams should include reference to refraction, the critical angle and total internal reflection.

Technician's notes

Be sure to check the latest safety notes on these resources before proceeding.

The following resources are needed for the class practical P3d.3a, per group:

plain A4 paper
semi-circular glass or acrylic block
ray box
power pack and leads
single slit
protractors

Take care. Ray boxes become hot after prolonged use.

Answers

Page 170 (top)

1. The angle of incidence at which the angle of refraction becomes equal to 90°.
2. 45.6°.
3. By repeated internal reflection.
4. It always hits the surface at an angle greater than the critical angle.
5. Infrared rays.
6. In endoscopes.

P3d.3a Critical angle

In this practical, you will be learning about the critical angle.

Apparatus

ray box

power pack and leads

single slit

semi-circular block

plain A4 paper

protractor

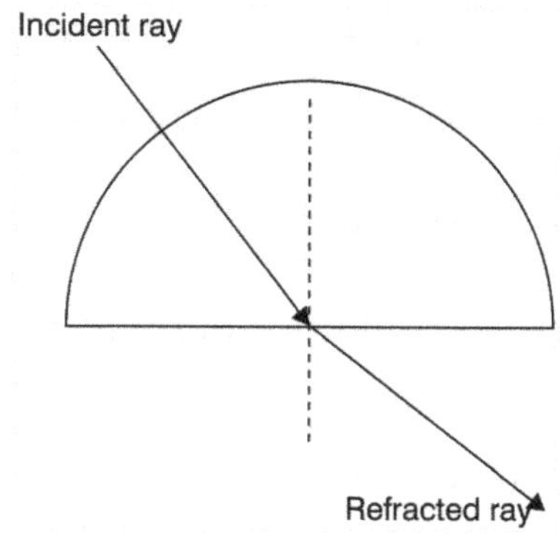

SAFETY INFORMATION

Take care when using glass blocks.

Ray boxes become hot after prolonged use.

Do not move around the laboratory in the dark.

Method

1. Place the semi-circular block in the centre of the paper and draw round it.
2. Draw a normal line to the block.
3. Send a ray of light into the block, to strike the intersection of the normal line and flat side of the block, as shown in the diagram above.
4. Increase the angle of the incident ray, as shown in the diagrams.

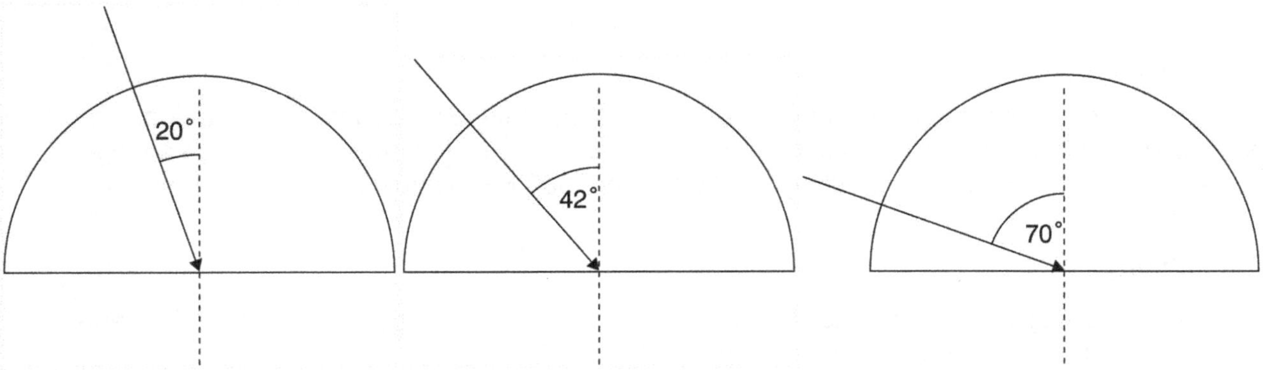

Analyse and interpret data

1. What is the critical angle for the block you used?
2. What is the angle of refraction at the critical angle?
3. What happens to the ray of light when the angle of incidence is greater than the critical angle?

Activity P3d.4 Properties of sound

Specification reference: P3b.3.23, P3b.3.24P

Learning objectives

- Describe the longitudinal nature of sound waves and how they can be reflected and refracted
- Know the range of frequencies for human hearing (20–20 000 Hz)

Learning outcomes

- Describe the production of sound by vibrating sources
- Describe the longitudinal nature of sound waves
- Show an understanding that a medium is needed to transmit sound waves
- Relate the loudness and pitch of sound waves to amplitude and frequency
- Describe how the reflection of sound may produce an echo
- Describe compression and rarefaction
- State that the range of human hearing is typically 20–20 000 Hz

Common misconceptions

Some students have difficulty in relating the longitudinal pressure variations of the air with the apparently transverse wave displayed on an oscilloscope screen.

Resources

Student Book pages 170–172

Worksheet P3d.4a Sound sources

Resources for class practical and demonstration (see Technician's notes, following)

Approach

Begin the activity

Use a loudspeaker, signal generator and candle flame to demonstrate low frequency longitudinal waves. Ask students to feel the vibrations of their throats as they hum a low note. Then place a ping-pong ball on a horizontal loudspeaker emitting a note at about 500 Hz to show that the cone is also vibrating.

Develop the activity

Discuss 'Sound waves' page 170 of the Student Book and answer the questions that follow. Students should then carry out the first part of the practical on Worksheet P3d.4a.

Then demonstrate the range of human hearing, introducing the ideas of amplitude and frequency at the same time. Show the effect of changing the amplitude and frequency of the sound from the signal generator.

Discuss 'Sounds humans can hear' on page 171 of the Student Book, including the Science in context feature on pages 171–172.

Students should carry out the second half of the practical on Worksheet P3d.4a. Students will need to ensure that they keep the string stretched so that vibrations can pass along it.

Bring the activity together

Ask students to work with a partner to produce a summary of the main points of the activity.

Technician's notes

Be sure to check the latest safety notes on these resources before proceeding.

The following resources are needed for the class demonstration P3d.4:

signal generator 10 Hz to 20 kHz
large loudspeaker with exposed cone
leads
candle
ping pong ball
clamp and stand to hold the lit candle about 20 cm in front of the loudspeaker

The following resources are needed for the class practical P3d.4a, per group:

tuning fork
large rubber bung
small plastic beaker
12 inch plastic rulers
long length of thick string
two empty plastic pots with small holes in their bases
two used matchsticks

Answers

Page 170 (bottom)

1. Longitudinal waves.
2. Small differences in air pressure.

P3d.4a Sound sources

You are going to investigate some of the properties of sound.

Apparatus

plastic ruler

plastic beaker

tuning fork

rubber bung

two plastic pots

two matchsticks

length of thick string

Method

Sound sources

1. Use your left hand to hold one end of the ruler so that the other end projects over the end of the table.
2. Use your right hand to twang the end of the ruler so that it vibrates.
3. Investigate how the pitch of the sound and the speed of the vibration depend on the length of the ruler over the edge of the table.
4. Hold the tuning fork by the handle. Sharply tap one of the two prongs on the rubber bung. Hold the tuning fork close to your ear and listen to the sound.
5. Fill the plastic beaker with water. Make the tuning fork emit a sound. Touch the surface of the water with one of the prongs. Explain your observations.

String telephone

1. Thread the string through both pots.
2. Tie a knot around a matchstick at each end of the string.
3. Stand away from your partner so that the string is stretched.
4. Place the pot against your ear. You should be able to clearly hear your partner when they talk into the open end of their pot.
5. Find out what happens if the string is allowed to go slack.

Activity P3d.5 Measuring the speed of sound

Specification reference: P3b.3.25P, P3b.3.26P, P3b.3.27P, P3b.3.28P, P3b.3.29P

Learning objectives

- Understand how an oscilloscope and microphone can be used to display a sound wave
- Relate the pitch of a sound to the frequency of vibration of the source
- Relate the loudness of a sound to the amplitude of vibration

Learning outcomes

- Describe an experiment to measure the speed of sound in air
- Describe an experiment using an oscilloscope to determine the frequency of a sound wave
- Be able to relate the pitch of a sound to its frequency and the loudness of a sound to its amplitude

Resources

Student Book pages 172–176

Worksheet P3d.5a Measuring the speed of sound

Resources for class practical and demonstration (see Technician's notes, following)

Approach

Begin the activity

Recap the two equations that lead to a value for the speed of a wave (speed = frequency × wavelength and speed = distance travelled ÷ time taken). Explain that in this lesson the students are going to consider one or both of these ideas to measure a value for the speed of sound.

Develop the activity

Introduce the method or methods chosen. Using a CRO with a microphone allows a sound pulse to be timed over a short distance. Use 'Measuring the speed of sound in air' on page 172 of the Student Book to help with this. Also, the worked example and investigative skills box will further reinforce understanding.

Calculate a value based upon the measurements obtained and consider the sources of error in the method chosen. Use this as an opportunity to consider experimental skills and discuss aspects such as reliability, accuracy and precision.

Discuss 'Using an oscilloscope and microphone to display sound waves' on page 174 of the Student Book. Using a microphone and oscilloscope, demonstrate the effect on the scope trace of varying the amplitude and frequency of the sound. Students need to understand that the wavy sound pictures are actually pressure–time graphs.

The students can attempt the developing investigative skills box on page 175. Also discuss 'Pitch, frequency and loudness' on page 176 of the Student Book and ask students to tackle the questions on page 174.

Bring the activity together

Set students the following problems.

- A sound moving through the ground has a frequency of 500 Hz and a wavelength of 4 m. How fast is it moving? [2000 m/s]

- The speed of sound in air is 340 m/s. What is the frequency of a sound which has a wavelength of 0.34 m? [1000 Hz]
- The speed of sound in a metal is 5000 m/s. What is the wavelength in the metal of a sound whose frequency is 10 kHz? [0.5 m]
- A sound of wavelength 2.0 m moves from air into a liquid. If the speed rises from 340 m/s to 3400 m/s, what happens to the frequency and the wavelength of the wave? [Frequency doesn't change as it is fixed by the source, wavelength drops to 0.2 m.]

Technician's notes

Be sure to check the latest safety notes on these resources before proceeding.

The following resources are needed for the class demonstration P3d.5:

stop clock and suitable area where echoes can be heard with a distance of about 70 m between the wall and the observer.
signal generator range 20 Hz to 20 kHz
large loudspeaker
microphone
clamp and stand to hold microphone in front of loudspeaker
oscilloscope with large display

The following resources are needed for the class practical P3d.5a, per group:

two microphones
fast recording device such as digital storage oscilloscope
sound source
measuring device
stop clock

Answers

Developing investigative skills, page 173–174

1. The students should repeat their measurements twice more and then take an average of the readings to minimise the errors in timing.

2. The distance is suitable because it is far enough that the student recording the echo should be able to distinguish the initial sound and the echo as separate sounds.

3. The sound makes eight round trips over a total distance of 800 m in 2.3 s so the speed of sound is $\text{speed} = \frac{\text{distance}}{\text{time}}$ = 347 m/s

4. Students could use two microphones a suitable distance apart, together with a digital oscilloscope. The students know the distance between the microphones and then they can record the time delay of a sound (a clap for example) of the sound reaching one microphone and then the other. The students could record the time delay over a series of different distances and then plot a graph of distance vs. time in which the gradient will be the speed of sound in air.

5. Using a graph helps improve the accuracy because it helps students to visualise any anomalous results as well as showing a trend for the data. In the case above, the graph enables the students to calculate the speed of sound directly using several data points.

6. A person's reaction time is around 0.3 seconds, so this would have the effect of making sound 'slower' than in reality. To avoid this error the student should increase the number of measurements and take an average, or use a computerised timing system instead.

7. The use of a data logger would eliminate human reaction time and improve timing accuracy of the sound.

Page 174

1. 20–20 000 Hz.

2. Ultrasound is sound above the range for human hearing.

3. Method should have the features of the Developing investigative skills box on measuring the speed of sound.

Developing investigative skills, page 175

1. Shield out as much background noise as possible.

2. and 3.

Tuning fork	**Number of waves**	**Time (s)**	**Frequency (Hz)**	**Note of tuning fork**
P	39	0.1	**390**	**G**
Q	22	0.1	**220**	**A**
R	25	0.1	**250**	**B or C**
S	33	0.1	**330**	**E**
T	24	0.1	**240**	**B?**
U	29 ½	0.1	**295**	**D**
V	35	0.1	**350**	**F**

4. The result for 'R' gives a frequency of 250 Hz which could indicate the tuning fork was B (246.94) or C (261.63). The number of wave cycles seems to be only approximate and generally measured to the nearest whole number of cycles.

5. The time sample could be reduced from 0.1 s. The number of wave cycles could be more accurately counted by using the squares on the oscilloscope. Results generally could be repeated and the mean taken. Results could be further repeated if they did not fit the pattern or they were anomalous.

P3d.5a Measuring the speed of sound

In this practical you will measure the speed of sound.

Apparatus

two microphones

fast recording device such as digital storage oscilloscope

sound source

measuring device

stop clock

SAFETY INFORMATION

Make sure that the sound source does not produce sound that is too loud as this will be uncomfortable for ears.

Method

The simplest method to measure the speed of sound in air uses two microphones and a fast recording device such as a digital storage oscilloscope.

1. Arrange a sound source and the two microphones in a straight line, with the sound source beyond the first microphone.
2. Measure the distance between the microphones (x). This is called microphone basis.
3. Measure the time of arrival between the signals (delay) reaching the different microphones (t).
4. Calculate speed of sound using speed = x / t

Analyse and interpret data

Distance x ________

Time t ________

Speed of sound ________

Evaluate data and methods

1. How could you check your measurements?
2. Would it be a good idea to repeat the timing part of the experiment? Explain your answer.

Activity P3d.6 Consolidation and summary

Learning objectives

- To review the learning points of the topic
- To test understanding through answering questions

Learning outcomes

- Be familiar with the knowledge and understanding summarised in the End of topic checklist
- Be able to apply this knowledge and understanding by answering End of topic questions

Resources

Student Book pages 177–182

Approach

Introduce the activity

Ask students to work with a partner to make a list of key words from this topic. They could then work together to produce a spider diagram showing how the different concepts are linked. They could compare their list with the list of key words given on page 177 of the Student Book. Discuss the checklist given on pages 177–178 and use questioning to see how much of the content students are comfortable with.

Develop the activity

Ask students to work individually to work through the End of topic questions on pages 179–182 of the Student Book without reference to the text. As they work, walk around the classroom observing their answers and questioning them as necessary to find out which questions are causing difficulties.

Bring the activity together

After a set period, ask the students to stop working and discuss any areas of difficulty you observed as you walked round the class. Students should complete any unanswered questions for homework, but you should stress that they should attempt the questions without reference to the text to see how much content they have retained.

Answers

End of topic questions mark scheme

Question	Correct answer	Marks
1 a)	in reflection light changes direction when it bounces off a surface, in refraction the light changes direction when it passes from one medium to another OR in refraction the angle of incidence does not equal the angle of refraction, as the ray of light bends towards or away from the normal	1
1 b) i)	diagram completed showing ray 1 going straight through, ray 2 refracted towards normal at first boundary and away from normal at second	1
1 b) ii)	angle of incidence	1
1 b) iii)	normal line	1
2 a)	you should draw a normal line to ensure that for each reflection the angles of incidence and reflection are the same	1
2 b)	one of: binoculars, bicycle reflector, cat's eyes, periscope	1
2 c) i)	reflection	1
2 c) ii)	total internal reflection and reflection	1
2 c) iii)	total internal reflection	1
3 a)	refractive index = sin 50° ÷ sin 21.7° = 2.07	1 1
3 b)	sin r = sin 90° ÷ 2.07 = 28.9°	1 1
4 a)	vibrations	1
4 b)	compressions and rarefactions travel through the air in the direction the wave is travelling	1 1 1
4 c)	sound waves cannot travel through a vacuum radio waves do not require a medium so they can travel through a vacuum	1 1 1

Question	Correct answer	Marks
5	150 ÷ 0.44 341 (or 340) m/s	1 1 1
6	85 m	1
7	the different densities of the hotter and cooler parts of the air cause the light to change direction as it passes through so when the light reaches you it has the ‘shimmering’ effect as some light arrives slight out of sequence from what you would expect	1 1 1
8	if the fibre is too wide, there are too many paths available to the light and so a pulse that starts as a ‘sharp’ pulse can become stretched out into a ‘smudge’ as the light following longer and shorter paths takes different times to get through the fibre	1 1 1
9	different wavelengths would be refracted different amounts so the signal would be split up	1 1
10	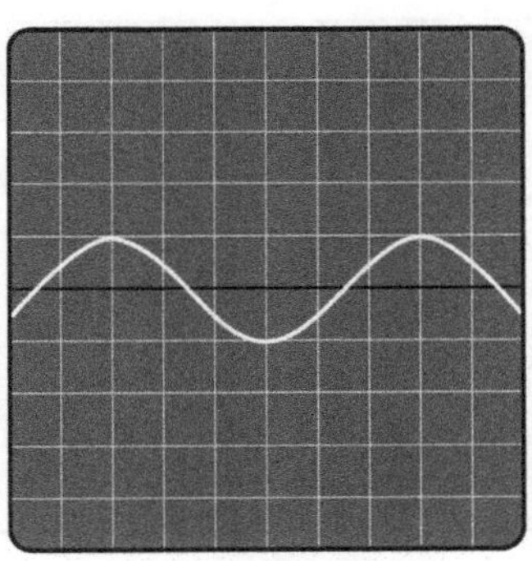 (a) 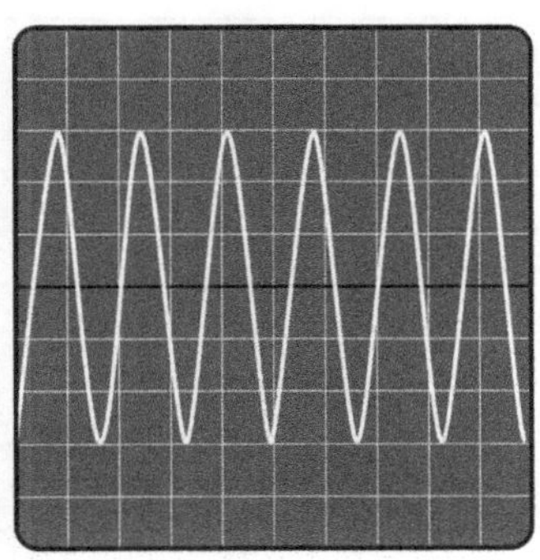 (b) 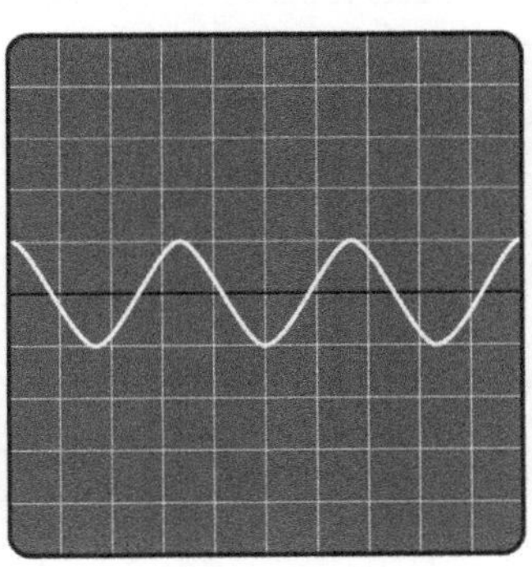 (c) 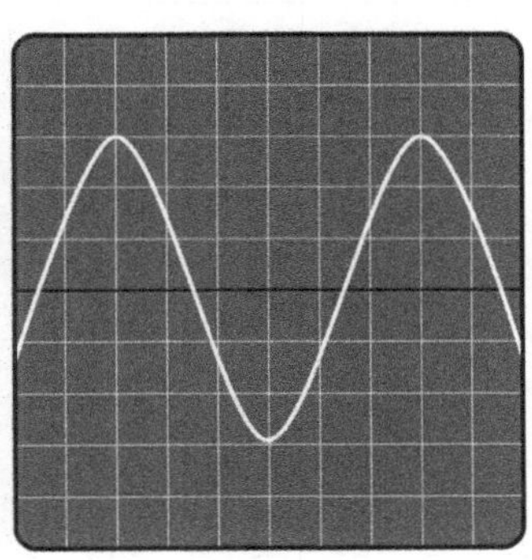 (d)	4 (1 mark for each)

Question	Correct answer	Marks
11 a) i)	the prongs of the tuning fork move back and forth	1
11 a) ii)	when the prongs move towards each other, the air is compressed between the prongs creating a compression, while a rarefaction occurs on the outside of the prongs	1
11 a) iii)	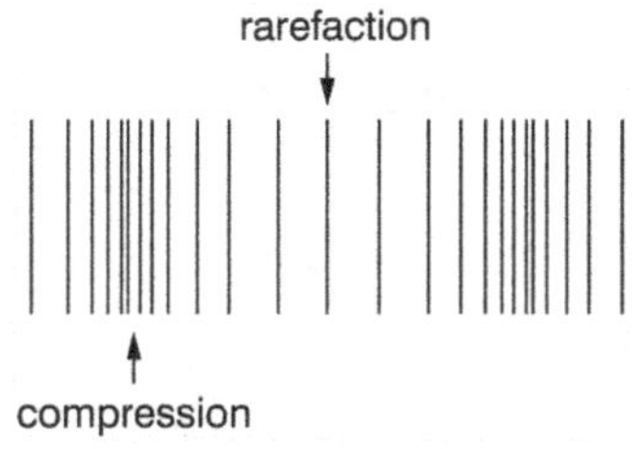	1
11 b)	when the tuning fork prongs are dipped into water ripples can be seen travelling out from the prongs	1 1
12 a)	refracted ray towards normal at first boundary then away from normal at second	1
12 b)	emergent ray at second glass/air boundary	1
12 c)	normal at 90 degrees to boundary (i.e. horizontal)	1
12 d)	depends on students' diagram	1
12 e)	light refracted when it enters a medium of different density to previous	1
13 a)	650 × 10–9/1.44 = 451 × 10–9 m	1 1
13 b)	shorter wavelength than red so towards violet end of spectrum	1 1
13 c)	6 .65 × 1014 Hz	1
	Total:	51

Section 3: Answers to exam-style questions

Question	Correct answer	Marks
2 a) i)	A	1
2 a) ii)	C	1
2 b) i)	suitable example, for example, light	1
2 b) ii)	transverse – vibration is perpendicular to direction of energy transfer longitudinal – vibration is in the same direction as energy transfer	1 1
2 c) i)	6 ÷ 10	1
2 c) ii)	1.7 (Hz)	1
2 c) iii)	speed = frequency × wavelength	1
2 c) iv)	1.7 × 0.05 0.085 (m/s)	1 1
3 a)	NOTE – any suitable method allowed (for example, ray box or sighting pins) suitable equipment how to carry out measurements plot graph of sin i against sin r	2 3 1
3 b)	total internal reflection (named) describe two conditions for law of reflection can bend fibre	1 2 1 1
	Total:	21

Section 4. Energy resources and energy transfers

Contents

Overview of the section

The first topic discusses the various stores of energy before students are reminded that energy is conserved. The relationship between efficiency, useful energy output and total energy input is introduced and used. The representation of energy transfers by Sankey diagrams is discussed and practised.

The transfer of energy by conduction, convection and radiation is discussed and the role of convection in everyday phenomena is explained. Finally, students learn about how insulation is used to reduce energy transfers from buildings and the human body.

The second topic is about work and power. Students learn the relationship between word done, force and distance moved, and between gravitational potential energy, mass, '*g*' and height. They also learn how to calculate kinetic energy. They learn how conservation of energy gives a link between gravitational potential energy, kinetic energy and work. Finally, they learn that power is the rate of doing work.

The final topic in this section covers energy resources and electricity generation, and the effect these have on the environment. The advantages and disadvantages of using various renewable and non-renewable energy resources to generate electricity are discussed.

Starting points

The Student Book provides a 'section opener', a double-page spread which sets the scene for students (see pages 186–187). It is structured in this way:

- an overview providing details of the areas of study
- six questions, the first three relating to aspects of physics that should be reasonably familiar to most students, and the final three relating directly to some of the topics to be studied
- a list of section contents.

The questions provide a structure for introducing the section.

1. What is the source of the energy store in your body?
2. Why are energy transfers never 100% efficient?
3. Name some energy stores.
4. What do you understand by the terms conduction, convection and radiation in relation to energy?
5. What is the principle of conservation of energy?
6. Name some ways in which electricity is generated. Which of these use renewable sources and which use non-renewable sources?

The section opener has two main purposes:

- to acknowledge the student's prior learning and to value it
- to provide a benchmark against which future learning can be compared.

The six questions or 'starting points' in the Student Book can be used in a number of different ways to introduce the section:

- You could ask students to consider the questions as an introductory homework task.
- You could put students into groups to share their own ideas and understanding and then report back to the whole class.
- Students could be given access to the Internet, preferably with a tight timescale, to find out the information required.

You could then use a spider chart or other form of wall chart to summarise everybody's ideas. Recording these initial ideas in this way allows you to retain them for reference as the individual topics are developed. In this way, your students' progress in learning can be readily acknowledged.

P4b Energy

Introduction

In this topic students will learn that energy is conserved and how it is transferred between various stores. From this, they will learn about the concept of energy efficiency and learn to show energy efficiency of devices using Sankey diagrams.

The topic concludes by considering how energy transfer takes place by conduction, convection and radiation. Based on this, students are shown how insulation is used to reduce energy transfers from buildings and the human body.

Links to other topics

Section	Essential background knowledge	Useful links
4 Energy resources and energy transfers		P4c Work and power P4d Human influences on the environment
5 Solids, liquids and gases		P5c Change of state P5d Ideal gas molecules

Topic overview

P4b.1	**Energy conservation and transfer** In this activity students learn that energy can be transferred from one store to another. They also learn that energy is conserved.
P4b.2	**Efficiency and Sankey diagrams** In this activity students learn that $\text{efficiency} = \frac{\text{useful energy output}}{\text{total energy input}}$ and how to use Sankey diagrams to show what happens to the energy that is input into devices.
P4b.3	**Conduction, convection and radiation** In this activity students learn how energy transfers take place by conduction, convection and radiation, and how this is applied to reduce energy transfers from buildings and the human body.
P4b.4	**Consolidation and summary** This activity provides an opportunity for a quick recap on the ideas encountered in the topic as well as time for the students to answer the End of topic questions in the Student Book.

Activity P4b.1 Energy conservation and transfer

Specification reference: P4b.4.2, P4b.4.3

Learning objectives

- Understand and be able to identify the following energy stores: chemical, kinetic, gravitational, elastic, thermal, magnetic, electrostatic, nuclear
- Identify and describe the following energy transfers: mechanical, electrical, by heating, by radiation (light and sound)
- Use the principle of conservation of energy

Learning outcomes

- Describe energy transfers involving energy stores
- Understand that energy is conserved

Common misconceptions

Students sometimes forget that some energy is converted to internal energy (heat) which is then lost to the surroundings. However, energy cannot be created or destroyed.

Students may also confuse the words 'conversion' and 'conservation'. Conversion of energy is changing it from one energy store to another (and some may be used to heat the surroundings) whereas conservation means that energy cannot be created or destroyed. The Student Book often uses the word 'transfer' instead of 'conversion'.

Resources

Student Book pages 189–193

Worksheet P4b.1a Energy

Resources for a demonstration (see Technician's notes, following). Practice this **before** the lesson. See CLEAPSS Recipe Book for more information.

Approach

Introduce the activity

Begin by asking students to state:

- what is needed to make a car move
- what is needed to make the hands of a wind-up clock move

The point is that energy is needed in both cases. In the car, the energy comes from the fuel, in the wind- up clock, the energy comes from a spring.

Ask students whether energy can be created or destroyed. [No]
What must be true about the total energy in to any system and the total energy out of any system? [the same]

Develop the activity

What kind of energy is in the spring in the clock? [potential energy] Use this as a springboard for discussing the various stores of energy set out on pages 189–190 of the Student Book. Ask students to work in pairs to produce a visual aid to help them remember the content of these pages.

Now move on to discuss energy conversions. Pages 191–193 of the Student Book can be used to guide the discussion. Make sure that the discussion covers the distinction between conversion and conservation as set out above.

Students should work in pairs to discuss and complete the questions on pages 192 and 193 of the Student Book to check their understanding of this activity.

They should then attempt Worksheet P4b.1a on their own. This worksheet is designed to summarise the content of this activity:

- Question 1 should be attempted by all students.
- Question 2 requires students to apply their knowledge of potential energy in a stretched band.
- Question 3 is a practical question which looks at energy changes when chemicals react.

You can demonstrate these changes using the instructions given in the Technician's notes below.

Bring the activity together

Discuss the students' answers to Worksheet P4b.1a. Go through any points that have caused difficulties and check that students understand why their answers differed from those given.

Ask students to choose an electrical item that they have at home, name it, and describe the energy conversions that take place when the item is used. For example, a radio converts electrical energy to sound energy, and there is also some heat energy around the radio when it is used.

Technician's notes

Be sure to check the latest safety notes on these resources before proceeding.

The following resources are needed for class demonstration P4b.1:

1 × 10 cm^3 and 2 × 100 cm^3 measuring cylinders
1 × 100 cm^3 and 2 × 250 cm^3 beakers
thermometer
stirring rod
Balance
concentrated (98%) sulfuric acid
ammonium nitrate
hydrochloric acid 1 mol/dm^3
sodium hydroxide solution 1 mol/dm^3

Splashproof goggles and nitrile gloves should be worn.

Concentrated sulfuric acid is extremely corrosive. It causes skin burns and eye damage. 1 M sodium hydroxide solution is corrosive to skin and eyes. Goggles must be worn.

Practice this **before** the lesson.

SAFETY INFORMATION

Students should wear goggles while watching this demonstration.

Demonstration should take place in a fume cupboard if possible. If not the laboratory must be well ventilated.

Method

a) Concentrated sulfuric acid with water

1. Measure about 1 cm^3 of concentrated (98%) sulfuric acid into a 10 cm^3 measuring cylinder and weigh.

2. Measure 100 cm^3 of ice-cold water with added ice cubes into a 250 cm^3 beaker. Record the temperature with a thermometer. Weigh the beaker, water and thermometer.

3. Add the sulfuric acid slowly while stirring the water gently with a polythene rod. Keep stirring until the temperature does not rise further. Record the temperature.

4. Weigh the 10 cm^3 measuring cylinder and calculate the mass of concentrated (98%) sulfuric acid.

5. Weigh the beaker, thermometer and reaction mixture.

6. Show that mass of reactants equals the mass of products.

b) Ammonium nitrate with water

1. Weigh 5 g of ammonium nitrate and put it into a 100 cm^3 beaker.

2. Measure 50 cm^3 of water into a 100 cm^3 measuring cylinder. Record the temperature with a thermometer.

3. Pour the water into the beaker containing the weighed sample of ammonium nitrate and stir with a polythene rod until the solid has dissolved. Record the temperature.

4. Weigh the beaker and solution of ammonium nitrate.

5. Show that mass of reactants equals the mass of products (55 g).

c) Hydrochloric acid and sodium hydroxide solution

1. Measure 50 cm^3 of 1 mol/dm^3 sodium hydroxide solution into a 250 cm^3 beaker.

2. Measure 50 cm^3 of 1 mol/dm^3 hydrochloric acid into a 100 cm^3 measuring cylinder.

3. Measure the temperature of the sodium hydroxide solution with a thermometer. Add the hydrochloric acid and stir. Record the highest temperature reached.

4. Show that mass of reactants equals the mass of products (for ease, use volumes).

Answers

Page 192

1. Gravitational – kinetic – gravitational.

2. In stretched or compressed springs.

3. a. Nuclear, b. chemical, c. kinetic, d. elastic, e. thermal, f. gravitational, g. electrostatic

4. a. Sound/sound wave, b. radiation of light/light wave, c. heating

Page 193

1. Energy cannot be created or destroyed.

2. Ball's gravitational store decreases, ball's kinetic store increases, thermal energy sore of surroundings increases; total amount of energy remains the same

3. a) Chemical energy store of fuel.

   b) Kinetic energy store of the car.

   c) Increased thermal energy store of the surroundings (and a small amount of energy transferred by sound).

P4b.1a Energy

Energy may be stored in various ways and moved from place to place (transferred) in various ways.

In the table below decide which names in the left column describe a way to store energy or a way to transfer energy.

Method

1. Put a tick in the correct box (two have been done for you already). Give one example in each case.

Type of energy	Store	Transfer	Example
chemical	✓		
electricity			
gravitational			
thermal			
heating		✓	
sound			
elastic			
light			
mechanical			

Making things happen

2. Imagine you make a catapult from two nails, a block of wood and an elastic band.

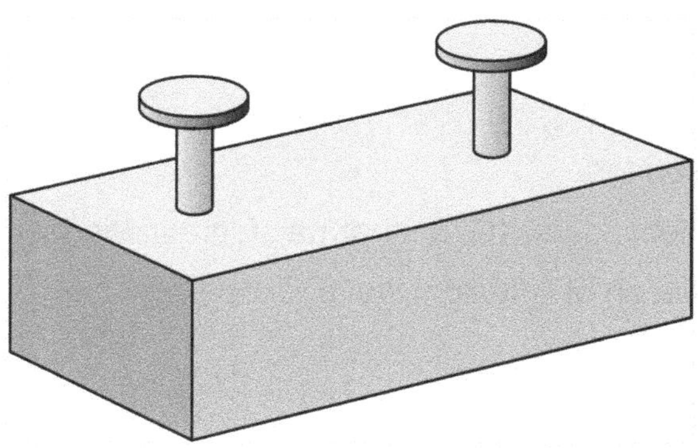

You use it to fire a small paper pellet. Now look at the three diagrams below. Each one (A, B and C) presents a firing position.

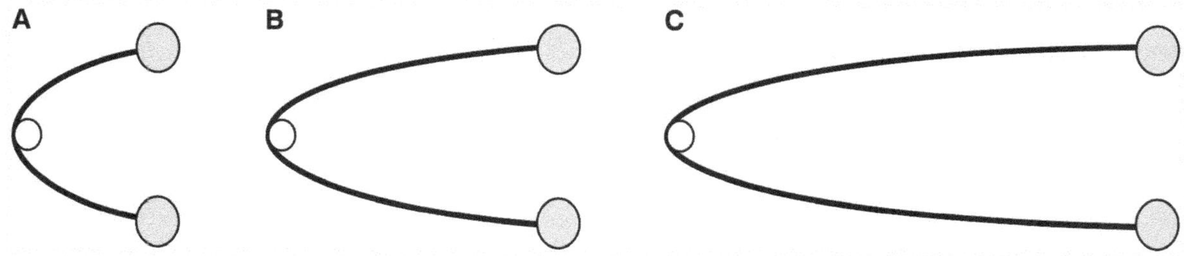

a) What does the stretched elastic band transfer its stored energy to when fired?

b) Which firing position will send the paper pellet the furthest?

c) In which firing position does the elastic band store the smallest amount of energy?

Energy from chemical changes

3. You will investigate energy and chemical changes. Record measurements made in the tables below. For the temperature change, put a '+' sign in front of the value if it is an increase in temperature and a '–' sign in front of the value if it is a decrease.

a) Concentrated sulfuric acid and water

Mass (g)			Temperature (°C)		
sulfuric acid	**water**	**mixture**	**before mixing**	**after mixing**	**change**

b) Ammonium nitrate and water

Mass (g)			Temperature (°C)		
ammonium nitrate	**water**	**mixture**	**before mixing**	**after mixing**	**change**

c) Reaction of hydrochloric acid and sodium hydroxide solution

Mass (g)			Temperature (°C)		
hydrochloric acid	**sodium hydroxide solution**	**mixture**	**before mixing**	**after mixing**	**change**

Analyse and interpret results

4. In experiments (a) and (c), chemical energy stored in compounds is transferred to the water. What happens to the temperature of the water?

5. In experiment (b), chemical energy stored in water is transferred to the ammonium nitrate. What happens to the temperature of the water?

Activity P4b.2 Efficiency and Sankey diagrams

Specification reference: P4b.4.4, P4b.4.5

Learning objectives

- Know and use the relationship $\text{efficiency} = \frac{\text{useful energy output}}{\text{total energy input}} \times 100\%$
- Describe a variety of everyday and scientific devices and situations, explaining the transfer of input energy in terms of the efficiency equation.
- Represent energy transfers by Sankey diagrams

Learning outcomes

- Be able to use the idea of efficiency and apply this to a range of energy transfers
- Be able to describe energy transfers using Sankey diagrams

Common misconceptions

Some students must be given plenty of support and encouragement when a lesson involves calculations. Lower-achieving students may have difficulty in using or drawing a scale diagram to keep account of energy transfers, but they are usually happy with the idea that energy in = total energy out.

Tracking down all of the outputs for a system can be tricky for some students, with some needing a lot of convincing that when energy appears to go missing, some heating of the environment happens.

Resources

Student Book pages 194–198

Worksheet P4b.2a Energy circus

Worksheet P4b.2b Heater efficiency

Resources for class practicals (see Technician's notes, following)

Approach

Introduce the activity

Remind students what they learnt in the previous activity about energy transfers and conservation. Ask students what they understand by the word 'efficiency'. Use some everyday live examples, such as a light bulb. Discuss how only some of the energy input is transferred as useful energy (light) but some is transferred as heat to the surroundings.

Display an EU rating chart from an electrical appliance such as a lamp, washing machine or car and use questioning to establish why increasing efficiency is a good thing to do in a world where energy is expensive and limited.

Develop the activity

Discuss Sankey diagrams and efficiency. Use Student Book pages 194–197 to guide the discussion. Stress that the diagram is drawn so that the thickness of each arrow is drawn to scale so that it shows the amount of energy it represents.

Students should then explore an 'energy circus' that you have set up around the room (see Technician's notes below for details). Give each student a copy of Worksheet P4b.2a to guide them. For each system, they have to identify the input and outputs. It may help to display on the board the names of the different energies involved in the 'circus'.

Students often fail to realise that for systems driven by batteries and humans the Sankey diagram has chemical energy as its input.

Students work in small groups to carry out the practical on Worksheet P4b.2b. If you do not have sufficient equipment for students to do this practical, then it could be done as a demonstration, as long as students are actively involved by being questioned about what they think will happen, what they observe, and being asked to explain the results. They should also carry out the calculation using the results in the 'analyse and interpret results' section.

Ask students to work through the Developing investigative skills feature on pages 194–195 of the Student Book and answer the questions as individuals. You may wish more able students to tackle the questions on Student Book page 198.

Bring the activity together

Discuss the students' answers to the Developing investigative skills feature. Revisit any answers that differ from those given and make sure that students understand why their answer is different to the model answer.

Discuss any practical activities undertaken during this lesson. Did the students encounter any difficulties? If so, how did they overcome them?

Technician's notes

Be sure to check the latest safety notes on these resources before proceeding.

The following resources are needed for the class practical P4b.2a, per group.

A range of energy transfer examples laid out around the laboratory. Could include the following:

lit candle – energy stored in chemicals to useful light and energy 'wasted' heating the surroundings (heat)
battery torch – energy stored in chemicals to useful heat and energy 'wasted' lighting up the surroundings (light)
mains fan – electrical to useful kinetic energy with energy 'wasted' making sound and heating the surroundings (sound and heat)
portable radio – energy stored in chemicals to useful sound and energy 'wasted' heating the surroundings (heat)
mains TV – electrical to useful light and sound and energy 'wasted' heating the surroundings (heat)
mains light – electrical to useful light and energy 'wasted' heating the surroundings (heat)
clockwork torch – energy stored in chemicals to useful light and energy 'wasted' heating the surroundings (heat)
hair dryer – electrical to useful heat and kinetic with energy 'wasted' making sound

The following resources are needed for the class practical P4b.2b, per group:

12 V 50 W electrical heater (inspected for damage around the cable entry points to avoid any trapped water boiling and pressurising inside the heater)
power pack (max 12 V)
clock

ammeter (0 to 5 A)
voltmeter (0 to 12 V at least)
three leads
250 cm^3 glass beaker
thermometer
clamp, stand and boss
access to balance and clocks
insulation materials (bubble wrap, cotton wool, cloth, paper …)
scissors and sticky tape

Students should not completely immerse the heater in the water. The cable connection must be above the water at all times.

Do not operate the heater when it is not in the water.

Answers

Developing investigative skills pages 194–195

1. Mass – balance. Length – (metre) ruler. Time – stopclock/stopwatch. Potential difference – voltmeter. Current – ammeter.

2. Since the timing involved starting and stopping the clock, the student probably decided that reaction time would make a difference, that is, recording times to 0.01 s was not realistic for hand timing. Could argue that to record to nearest 0.1 s is still quite ambitious.

3. Table completed correctly for values given.

4. Graph plotted correctly for data calculated.

5. Suitable sentence or two that correctly describes the pattern in the data.

6. Need to realise that the gradient of the line is equal to the efficiency.

7. Sensible comment required – could suggest extending the range, for example, to 0.11 kg to see if the pattern continues, or could suggest testing intermediate values, for example, 0.04 kg, to add detail within the range already tested.

Page 198 (top)

1. How much energy is transferred each way.

2. Sankey diagram should have 1000 J as input. There should be two output arrows, one labelled 400 J useful heat energy, the other 600 J waster heat energy. The sizes of the arrows should be in proportion.

3. 40%

4. a) 39 200 J

   b) 2%

Page 198 (bottom)

Student's own Sankey diagrams.

P4b.2a Energy circus

In this activity you will explore different energy transfer systems that are located around the classroom.

Method

For each energy transfer system:

1. Describe what the system does.
2. Identify the input, useful output and waste energy transfers.

Analyse and interpret data

Use the information provided about each of these systems to draw a labelled scale drawing of a Sankey diagram to represent the energy flows through them.

Name each type of energy as well as quantify it.

1. An incandescent lamp takes in 100 W from the mains, of which 90 W is wasted in heating the environment.
2. For every 20 J of chemical energy that transfers into a gas cooker, only 19 J is transferred to useful heat energy.
3. A portable MP3 player transfers 1 J to useful sound energy, but wastes another 2 J along the way.
4. A fired gun transfers 800 J of kinetic energy to a bullet, with 10 J of sound energy and 140 J of heat energy.

P4b.2b Heater efficiency

In this practical activity you will calculate the efficiency of a heater.

Apparatus

power pack

electrical heater

ammeter

voltmeter

thermometer

three leads

glass beaker

clock

insulation

scissors and sticky tape

balance

clamp, stand and boss

SAFETY INFORMATION

DO NOT completely immerse the heater in the water.

The cable connection must be above the water at all times.

Do not operate the heater when it is not in the water.

Tell the students to turn the power pack on with the heater out of the water.

Method

1. Connect the ammeter and heater in series with the power pack as shown.

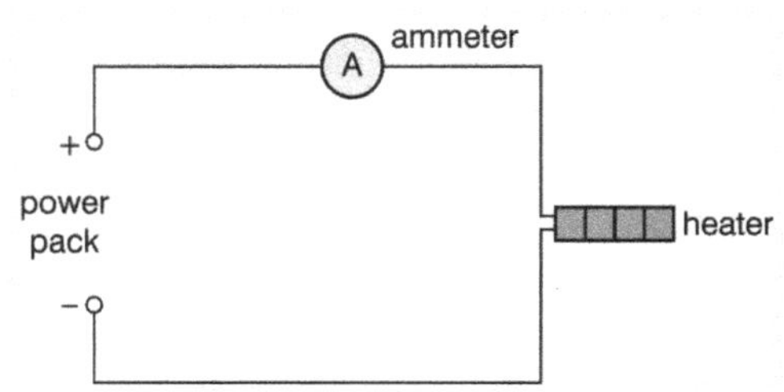

2. Turn the power pack to zero volts.

3. Connect the voltmeter across the power pack as shown. You should not need to change the circuit that you have already built.

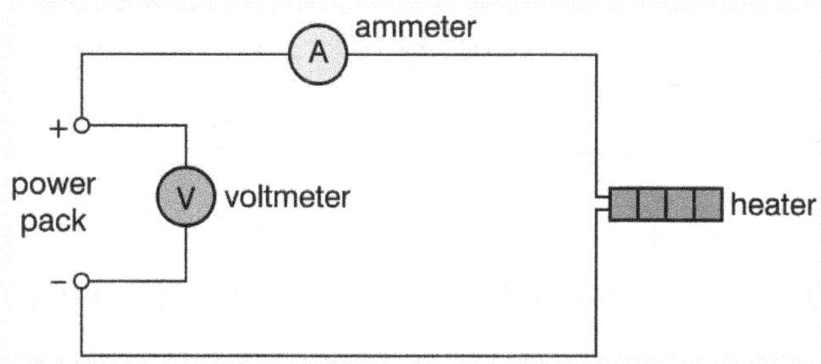

4. Fill the beaker three-quarters full of cold water. Find its mass in kilograms.

5. Clamp the heater in place so that about three-quarters of its length is under the water. Make sure the cable connection is out of the water.

6. Switch on the power pack and slowly turn up the voltage. Check that the ammeter shows the voltage, then switch **off** again.

7. Switch on the power pack and slowly turn up the voltage. Check that the voltmeter shows the voltage, then switch **off** again.

8. Note the temperature of the water.

9. Start the clock as you turn on the power pack and raise the voltage to about 12 V.

10. Note the reading of the ammeter and voltmeter.

11. Switch off the power pack after exactly five minutes.

12. Stir the water. Note its temperature once the thermometer has settled to its final value.

If you have time, you could repeat the experiment with fresh cold water and lots of insulation all around the beaker.

Analyse and interpret data

Calculate the efficiency of the system.

Use your results to complete these calculations.

The total energy transferred in (J) = voltage (V) × current (A) × time (s) Total energy transferred in =___V × _____A × 300 s =___J Initial temperature =_°C

Final temperature =_°C

Temperature rise =________°C –___°C =___°C Mass of water =___kg

Useful energy transferred to water (J) = mass (kg) × temperature rise (°C) × 4200 J/kg/°C

Useful energy transferred (J) =_____kg ×___°C × 4200 J/kg/°C

Useful energy transferred out =____J

$$\text{efficiency} = \frac{\text{useful energy out}}{\text{total energy in}} = \_\_\_\_\_\%$$

Total energy transferred in = ________ V x ________ A x 300 s = ________ J

Initial temperature = ________ °C

Final temperature = ________ °C

Temperature rise = ________ °C - ________ °C == ________ °C

Mass of Water = ________ kg

Useful energy transferred to water (J) = mass (kg) x temperature rise (°C) x 4200 J/kg/°C

Useful energy transferred (J) = ________ kg x ________ °C x 4200 J/kg/°C

Useful energy transferred out = ________ J

$$\text{efficiency} = \frac{\text{useful energy out}}{\text{total energy in}} = \_\_\_\_\_\_\_\_ \%$$

Activity P4b.3 Conduction, convection and radiation

Specification reference: P4b.4.6, P4b.4.7, P4b.4.8, P4b.4.9, P4b.4.10

Learning objectives

- Describe how thermal energy transfer may take place by conduction, convection and radiation
- Explain the role of convection in everyday phenomena
- Explain how emission and absorption of radiation are related to surface and temperature
- Explain how insulation is used to reduce energy transfers from buildings and the human body

Learning outcomes

- Be able to describe mechanisms of thermal energy transfer
- Understand how knowledge of energy transfer mechanisms is used to reduce energy transfers in a range of situations

Common misconceptions

Conduction and convection sound very similar so are often confused. Many students have difficulty in accepting that a conductor and an insulator can both be at the same temperature, yet one feels colder than the other. Take time to explain this in terms of energy transfer.

Many students have difficulty with the idea that continually transferring energy into something can result in a steady temperature.

Resources

Student Book pages 198–208

Worksheet P4b.3a Insulating a house

Resources for demonstrations and class practical (see Technician's notes, following)

Approach

Introduce the activity

Ask students what they understand by the terms 'conduction' and 'convection'. They may have ideas about conduction from previous learning about electricity. Ask them how the term applies to thermal energy.

Set up a Bunsen burner on the front bench with a yellow flame. Introduce the idea of convection currents. Use a metal windmill to show that there is an upwards movement of gas from the flame.

Develop the activity

Discuss conduction and convection using pages 198–200 of the Student Book. Students should attempt the questions within the text with a partner to check understanding. Demonstrating convection can either be done as a demonstration or students can do it in small groups, depending on the availability of equipment. Condensation on the outside of the beaker can obscure the view of the convection currents.

You may wish more able students to attempt the Extension feature on page 200 of the Student Book. For details of apparatus that will be needed, see Technician's notes below.

Show students samples of different materials. Ask them to say whether they think they are conductors or insulators.

Move on to discuss radiation using Student Book pages 200–201 as a guide. If possible, set up a Leslie's cube and an infrared heat sensor with a large display meter. Show them that all four walls have the same thickness and so should transfer energy from hot water inside at the same rate. Then fill the cube with boiled water from the kettle and show that the black surface emits much more radiation than the white one. Students should attempt the questions within the text on page 202 to check understanding.

Finally, discuss everyday examples of thermal transfer using Student Book pages 202–204 as a guide. If possible, students could carry out the practical on Worksheet P4b.3a to demonstrate the principle of insulating a house.

Bring the activity together

Ask students to write down their own definitions of conduction and convection. Place three items of clothing on the bench (pullover, blanket, rubber soled shoes). Use questioning to elicit from the students that the pullover reduces convection from the skin, the blanket reduces radiation and the rubber soles of the shoes reduce conduction.

There are opportunities for students to understand that the air trapped in the pullover reduces conduction, leading to a low surface temperature which in turn reduces radiation.

Give each group with a sheet of paper. Ask them to design a house which will keep cool inside in hot weather – without the use of air conditioning. Give the groups five minutes before sharing their ideas with the whole class. You could look for thick insulated walls and roofs, small windows, white outside walls etc.

Technician's notes

Be sure to check the latest safety notes on these resources before proceeding.

The following resources are needed for the class demonstrations P4b.3.

Convection demonstration

250 cm^3 glass beaker
Bunsen burner, heatproof mat, tripod and gauze
forceps
metal windmill
hollow glass tube long and wide enough to deliver a potassium manganate(VII) crystal to base of beaker; a funnel placed into the top of the tube makes it easier to get the crystal into the tube
potassium manganate(VII) crystal
eye protection

Potassium manganate(VII) is oxidising, harmful and dangerous for the environment.

Wear eye protection.

Leslie's cube demonstration

Leslie's cube and hot water
infrared heat sensor and large display meter

The following resources will be needed for the extension activity on page 200 of the Student Book (per group):

copper strip	Bunsen burner, heatproof mat and tripod
petroleum jelly	stop clock
pins	eye protection

The following resources are needed for the class practical P4b.3a, per group:

ray box (with 12 V tungsten filament lamp)
power pack (max 12 V)
cardboard box (shoe box)
thermometer with rubber sleeve
thick slab of fluffy material to cover the top of the model house – a thick piece of expanded polystyrene, thick carpet, cotton wool or bubble wrap, for example.
access to more fluffy material to stick on the walls of the house
sticky tape to hold insulation in place

Answers

Page 199

1. They contain electrons that can move freely and transfer energy.
2. There are no particles to transfer energy by colliding with each other.
3. Vibrations of the particles are passed on through the bonds between the particles.
4. Some wax is put on one end of a metal rod. The other end of the metal rod is heated until the wax on the other end melts.

Page 200 (top)

Use petroleum jelly to attach drawing pins at regular distances along the copper strip. Heat one end of the strip and measure the time it takes for each drawing pin to fall off. Plot a graph of time until the drawing pin falls off against distance from the point that is being heated.

Page 200 (bottom)

1. The particles are free to move.
2. Warm air expands, which makes it less dense. Less dense air floats up above more dense (cooler) air.
3. Heat some potassium manganate(VII) crystals in water.
4. Fibres in the insulation create air pockets. This restricts the movement of the air and so convection currents cannot form.

Page 202

1. It can travel through a vacuum.
2. A hot object.
3. Temperature and type of surface.
4. The dull, black side.

Developing investigative skills, page 204

1. A cooler room temperature will increase the cooling rate so it is useful to know this temperature to make firm conclusions.
2. Temperatures can become uneven in a beaker of still water. Often warmer at the top and cooler below. Stirring mixes up the particles to ensure an even temperature.
3. Every 3 minutes.

4. a) A smoother curve.

   b) A less smooth curve.

   c) Metal is a good conductor so increasing energy transfer from the water which would cool quicker than on wood.

5. Cooling rate from 90 °C to 70 °C is 20 °C ÷ 9 minutes = 2.2 °C loss per minute.

Cooling rate from 70 °C to 50 °C is 20 °C ÷ 18 minutes = 1.1 °C loss per minute.

6. Insulate the beaker and use a lid. Increase room temperature.

7. Record the cooling curve for different starting temperatures.

Page 205

1. Convection.

2. The top of the room.

3. a) Conduction – reduced by having a vacuum (no particles to connect).

   b) Convection – vacuum space has no particles to form convection currents.

   c) Radiation – silver surfaces reflect the energy and do not absorb or emit it.

4. The flask reduces energy transfer in both directions, so hot drinks do not lose their energy and cold drinks are not warmed by energy entering from outside.

Developing investigative skills, pages 207–208

1. Sensible and constructive safety precautions such as: stand up when pouring hot water, keep desk tidy and clutter free, do not carry hot water across the room, have the kettle and experiment on the same desk.

2. It will illustrate the effect of glass on its own rather than glass and another material.

3. Each line plotted correctly with a smooth curve.

4. Black paper = cools fastest; glass only = 2nd fastest; insulation with air bubbles = 3rd fastest; silver foil one cools slowest.

5. Temperature difference (between water and room temperature) is greatest in the first 5 minutes. Therefore, energy transfer is highest.

6. Shiny foil reflects radiated heat back in to the water. Hot shiny foil is also a poor emitter of radiation / heat.

7. Black paper is a good absorber of energy radiated from the hot water. The black paper emits lots of radiation to the air.

8. 20 °C, but allow 19 °C.

9. Stir the water a few seconds before each measurement. The thermometers will need lifting from the container to read them. Alternatively, a temperature sensor with a data logger could be used.

10. All curves less steep. All curves (eventually) plateau at 30 °C.

Page 208

1. More than half the energy wasted is through these two features.

2. Answer should refer to trapped air in the walls, roof, windows, etc.

3. Air is a bad conductor (good insulator) so keeping a layer near the body reduces heat loss by conduction. Keeping the layer trapped reduces heat loss by convection.

4. In hotter climates, you may want to lose heat, so loose clothing allows the air to move and transfer heat away.

P4b.3a Insulating a house

In this practical you are going to look at the effect of insulation on heat loss from a model house.

Apparatus

ray box with 12 V lamp

cardboard box to cover ray box

thermometer with rubber sleeve

thick slab of insulation

thin insulation

sticky tape

power pack

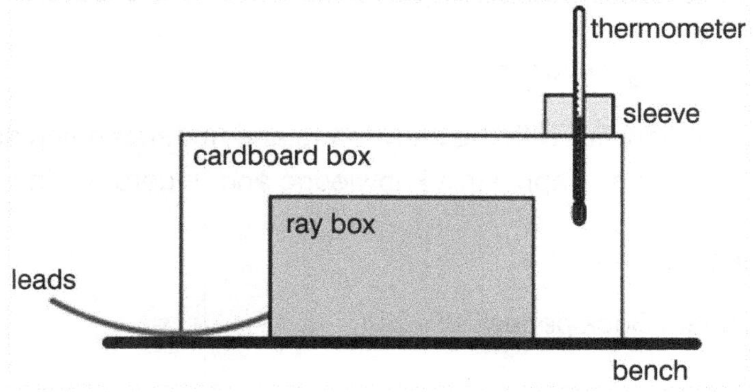

SAFETY INFORMATION

Ray boxes become hot when they have been switched on for a while.

Method

1. Set up your model house as shown in the diagram above.
2. Put the ray box (the heater) on the bench.
3. Connect it to the power pack, turn up the voltage to full and switch it on.
4. Put the cardboard box (the house) on top of the power pack.
5. Poke the thermometer through the hole in the top of the house.
6. Leave the house to reach a steady temperature.
7. Now add a layer of fluffy insulation on the top of the house.
8. Note what effect this has on the new steady temperature of the house.

Analyse and interpret data

1. How could you adjust this investigation to look at different factors which affect the rate of heat loss from a house?
2. What did you notice about the temperature reached inside the model house with and without the insulation?
3. Explain your results in terms of convection, conduction and radiation.

Activity P4b.4 Consolidation and summary

Learning objectives

- To review the learning points of the topic
- To test understanding through answering questions

Learning outcomes

- Be familiar with the knowledge and understanding summarised in the End of topic checklist
- Be able to apply this knowledge and understanding by answering End of topic questions

Resources

Student Book pages 209–211

Approach

Introduce the activity

Ask students to work with a partner to make a list of key words from this topic. They could then work together to produce a spider diagram showing how the different concepts are linked. They could compare their list with the list of key words given on page 209 of the Student Book. Discuss the checklist given on page 209 and use questioning to see how much of the content students are comfortable with.

Develop the activity

Ask students to work individually to work through the End of topic questions on pages 210–211 of the Student Book without reference to the text. As they work, walk around the classroom observing their answers and questioning them as necessary to find out which questions are causing difficulties.

Bring the activity together

After a set period, ask the students to stop working and discuss any areas of difficulty you observed as you walked round the class. Students should complete any unanswered questions for homework, but you should stress that they should attempt the questions without reference to the text to see how much content they have retained.

Answers

End of topic questions mark scheme

<table>
<tr><th>Question</th><th>Correct answer</th><th>Marks</th></tr>
<tr><td>1</td><td>the thin layers will trap air between them
air is an insulator
and so will reduce thermal transfer from the body</td><td>1
1
1</td></tr>
<tr><td>2</td><td>energy from the hot water
is transferred from the inner to the outer surface of the metal by conduction
energy is transferred through the still air by radiation</td><td>1
1
1</td></tr>
<tr><td>3 a)</td><td>3500
J</td><td>1
1</td></tr>
<tr><td>3 b)</td><td>70%</td><td>2</td></tr>
<tr><td>4 a)</td><td>20
J</td><td>1
1</td></tr>
<tr><td>4 b)</td><td>40%</td><td>2</td></tr>
<tr><td>5</td><td>completed table as shown
<table>
<tr><th>Object</th><th>Input energy</th><th>Useful energy output</th><th>Wasted energy</th><th>Efficiency</th></tr>
<tr><td>light bulb</td><td>100 J</td><td>10 J</td><td>90 J</td><td>10%</td></tr>
<tr><td>torch</td><td>70 J</td><td>55 J</td><td>15 J</td><td>79%</td></tr>
<tr><td>radio</td><td>250 J</td><td>210 J</td><td>40 J</td><td>84%</td></tr>
</table></td><td>12</td></tr>
<tr><td>6</td><td>liquids and gases become less
dense
when heated</td><td>1
1
1</td></tr>
<tr><td>7</td><td>the particles in a solid are fixed in position
so vibrations pass readily
from particle to particle</td><td>1
1
1</td></tr>
<tr><td>8</td><td>air above the convection heat absorbs energy
becomes less dense and rises, being replaced by cooler air
which in turn is warmed and rises so a convection current transfers energy from the convection heater around the room</td><td>1
1
1</td></tr>
<tr><td>9</td><td>serving dishes are made of insulators
so that they do not transfer energy away from (or into) the food
this keeps hot food hot (or cold food cold) and makes it less likely that someone handling the dish will be burned</td><td>1
1
1</td></tr>
<tr><td>10</td><td>any three of the following: the material the object is made from:
the volume and surface area of the object
the appearance of the surface (light, dark, dull, shiny)
the temperature difference between the object and the surroundings</td><td>3
(1 mark for each)</td></tr>
</table>

Question	Correct answer	Marks
11	the useful energy output arrow for the Sankey diagram should be ¼ the width of the input energy arrow	1 1 1
12	lubricants improve the efficiency of an engine because they reduce friction so reduce the amount of energy 'wasted' by being transferred as heat to the surroundings as work is done to overcome frictional forces	1 1
13	65%	2
14	280 J	1 1
15	480 J	1 1
16	9.8 J	1 1
17	petrol engine 30% efficiency steam–diesel hybrid 25% petrol engine is more efficient	1 1 1
18	This is a level of response question. For example: Level 1: Particles in warm air move around quickly and collide with inner particles in glass OR particles in air are moving around quickly Level 2: Particles of warm air move around and collide with particles in (inner) glass. These glass particles move / vibrate quicker and pass vibrations through the glass OR vibrations / kinetic energy of particles passed through glass Level 3: Particles of warm air move around and collide with particles in (inner) glass. These gain energy and vibrate more. Vibrations / kinetic energy of particles passed through glass. Air particles on the outside collide with vibrating.(outer) glass particles and bounce off with more kinetic energy OR kinetic energy is passed on by particle collisions, from particles of the warm air on the inside to (inner) glass particles, through vibrations to (outer) glass particles and then to the air particles on the outside	 2 1 4 3 6 5
	Total:	78

P4c Work and power

Introduction

In this topic students learn the definition of 'work' in physics. They then learn about gravitational potential energy, kinetic energy and how the principle of conservation of energy provides a link between gravitational potential energy, kinetic energy and work. Finally, the topic considers power as the rate of transfer of energy or the rate of doing work.

Links to other topics

Section	Essential background knowledge	Useful links
1 Forces and motion		P1c Forces, movement, shape and momentum
4 Energy resources and energy transfers	P4b Energy	P4d Energy resources and electricity generation

Topic overview

P4c.1	**Work** In this activity students will learn the relationship between work, force and distance moved in the direction of the force. They will appreciate that work done is equal to energy transferred.
P4c.2	**Gravitational potential energy, kinetic energy and conservation of energy** In this activity students learn how to calculate gravitational potential energy, kinetic energy and how the conservation of energy relates the two types of energy.
P4c.3	**Power** In this activity students learn the relationship between power, work done and time taken.
P4c.4	**Consolidation and summary** This activity provides an opportunity for a quick recap on the ideas encountered in the topic as well as time for the students to answer the End of topic questions in the Student Book.

Activity P4c.1 Work

Specification reference: P4c.4.11, P4c.4.12

Learning objectives

- Know and use the relationship between work, force and distance moved in the direction of the force:
  work done = force × distance
  $W = F \times d$
- Understand that work done is equal to energy transferred

Learning outcomes

- Be able to use the equation work done = force × distance
- Know the link between forces and energy transfer

Common misconceptions

'Work' has a special and precise meaning in physics. Students need to realise that for work to be done, the distance moved has to be in the same direction as the force and that both force and distance are required for an energy transfer to take place.

In order to evaluate the results of the practical, students need to appreciate that energy transferred by friction can reasonably account for the differences in the work done and the energy gained.

Resources

Student Book pages 212–214

Worksheet P4c.1a Doing work

Resources for class practical and demonstration (see Technician's notes, following)

Approach

Introduce the activity

Invite a student to stand against a wall and push very hard against it. Explain to the class that although the student is getting tired, they aren't doing any work because there is no energy transfer to the wall.

Ask for a volunteer. Both of you need to wear eye protection. Hold one end of a long length of rubber tubing and invite the student to take hold of the other end and extend it as far as they can. Ask if an energy transfer has taken place. Explain that the student has applied a force and moved against it so work has been done. Be careful to release the elastic potential energy safely.

Discuss 'work' and explain that it has one meaning in physics and wide range of different ones in everyday life.

Develop the activity

Discuss the concept of 'work' as described on pages 212–213 of the Student Book. You could set the worked examples for the students to attempt in pairs without reference to the given solution, and then discuss the given solution and how it differs from the students' solutions.

If possible, the practical on Worksheet P4c.1a should be carried out in small groups, but if there is not enough equipment for this, then it can be carried out as a demonstration as long as students are actively involved through questioning about what they think will happen, their observations, and analysis of the results.

The practical work is a useful test of the idea of energy conservation. Within the random and systematic errors prevalent in practical work at this level, the work done pulling the trolley along the inclined track and straight up into the air should be the same, as friction should be negligible. The string is necessary because the force meter only works correctly when it is vertical.

You may then wish more able students to attempt the extension feature on page 214 of the Student Book.

Bring the activity together

Ask students to write down a definition of 'work' as used in physics. You can use the questions on page 214 of the Student Book to check understanding.

Technician's notes

Be sure to check the latest safety notes on these resources before proceeding.

The following resources are needed for the class demonstration P4c.1:

eye protection
rubber tubing

The following resources are needed for the class practical P4c.1a, per group:

dynamics trolley
trolley track
clamp and stand to raise one end of track
metre rule
force meter (0–10 N)
pulley wheel on end of track
strong thread or string

Students will measure the force required to pull the trolley along the track when one end has been raised above the bench and compare this with the force required to pull the trolley directly up into the air.

Answers

Page 214 (top)

1. 250 J.
2. 500 N.
3. 80 J.
4. 50 m.
5. 400 N.

Page 214 (bottom)

1. $8.45 \times 10^{12}/8 \times 10^{3} = 1.06 \times 10^{9}$ N.
2. Series of manoeuvres used before landing to get rid of excess energy.
3. 1650 °C. The orbiter is covered with ceramic insulating materials designed to protect it from this heat. The materials include:

Reinforced carbon-carbon (RCC) on the wing surfaces and underside.

High-temperature black surface insulation tiles on the upper forward fuselage and around the windows. White Nomex blankets on the upper payload bay doors, portions of the upper wing and mid/aft fuselage. Low-temperature white surface tiles on the remaining areas.

P4c.1a Doing work

You are going to investigate the work done pulling a trolley up a ramp (or inclined track).

Apparatus

trolley

force meter

track

pulley

clamp and stand

metre rule

string

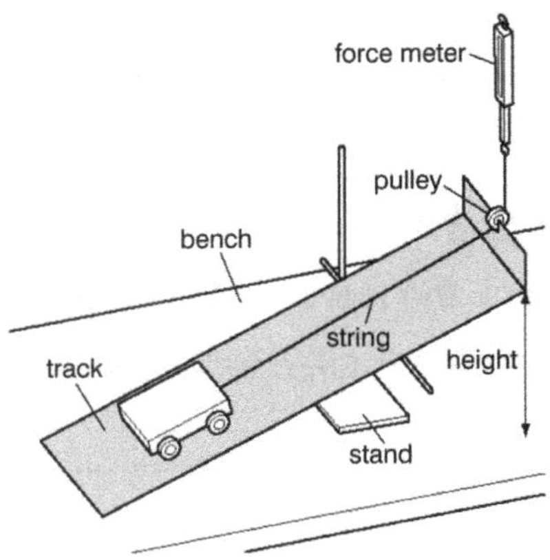

SAFETY INFORMATION

Ensure knots are tightly tied to prevent trolley dropping onto the floor or someone's foot.

Method

1. Measure the length of the track and set it up on the bench as shown above.
2. Arrange the clamp so that the lower end of the top of the track is 5 cm above the bench.
3. Check that the force meter is zeroed correctly.
4. Suspend the trolley directly from the force meter. Note the weight of the trolley.
5. Slowly pull the trolley up the ramp with the string and pulley as shown. Note the reading on the force meter, the force pulling the trolley in its direction of motion.
6. Repeat step 5 for the different ramp heights given in the table below.

Recording your results

1. Record your results in a table like this one.

Trolley weight (kg)	Ramp height (m)	GPE gained (J)	String force (N)	Ramp length (m)	Work done (J)
	0.05				
	0.10				
	0.15				
	0.20				
	0.25				
	0.30				

Analyse and interpret data

1. Use the trolley weight and ramp height to calculate the gravitational potential energy gained. Make sure that you have quantities in the correct units.
2. Use the string force and the ramp length to calculate the work done.
3. Comment on the values of work done and gravitational potential energy gained for the different settings of the ramp.

Activity P4c.2 Gravitational potential energy, kinetic energy and conservation of energy

Specification reference: P4c.4.13, P4c.4.14, P4c.4.15

Learning objectives

- Know and use the relationship between gravitational potential energy, mass, gravitational field strength and height: gravitational potential energy = mass × gravitational field strength × height
  GPE = $m \times g \times h$
- Know and use the relationship: kinetic energy $= \frac{1}{2} \times \text{mass} \times \text{speed}^2$

  KE = $\frac{1}{2} \times m \times v^2$
- Understand how conservation of energy produces a link between gravitational potential energy, kinetic energy and work

Learning outcomes

- Be able to calculate the energy transferred in number of situations

Common misconceptions

Although most students readily accept that moving objects have kinetic energy, its relationship with the mass and velocity of the object can be difficult to grasp.

Not only does this activity introduce two different equations, but also one of them is arguably the most complex one of the entire International GCSE Physics course. Students who are not mathematically inclined will need to take it slowly.

Resources

Student Book pages 215–218

Practical P4c.2a Up and down

Practical P4c.2b Pendulum swing

Resources for demonstration and class practicals (see Technician's notes, following)

Resources for final demonstration: a golf ball

Approach

Introduce the activity

Set up a sand-filled bottle tied to a metre-long piece of string – clamp the string to a bench so that the bottle can swing freely. Use class discussion to establish the transfers between KE and GPE for the pendulum as it swings. It is not at all obvious to students that a swinging mass has zero kinetic energy at the extremities of its swing.

You could also carry out this activity as a class practical – instructions are given on Worksheet P4c.2a. Run a dynamics trolley down a track set up at an angle. Arrange a second track so that the trolley passes smoothly onto it and carries on moving horizontally. Point out the energy transfers between KE and GPE. Ignore the role of friction unless your class bring it to your attention.

Develop the activity

Discuss the content of Student Book pages 215–216. When using the worked examples, follow the approach suggested in Activity P4c.1. You may wish more able students to look at the extension feature on page 218 of the Student Book. Students can then look at the questions on page 216 to check understanding of the content of the lesson so far.

Discuss 'Energy at work' on pages 216–217 of the Student book. Again, set the worked example for students to tackle in pairs and then discuss how the given solution and students' solutions differ and why.

Worksheet P4c.2b may be used at this point. If resources allow, students may do this practical in small groups, or you may do it as a demonstration, with students actively involved as described above.

Bring the activity together

Allow students to tackle the questions on page 218 of the Student Book in pairs to check understanding. Then raise a golf ball up from the floor to above your head. Then drop it and let it bounce until it stops moving. Use class questioning to establish all the energy transfers.

Technician's notes

Be sure to check the latest safety notes on these resources before proceeding.

The following resources are needed for the first demonstration P4c.2 (or per group if carried out as practical P4c.2a):

sand-filled plastic bottle with string through the lid
G-clamp

The following resources are needed for the second demonstration (dynamics trolley):

dynamics trolley
two trolley tracks
block to raise one end of one of the tracks

The following resources are needed for the class practical P4c.2b, per group:

light gate
computer running light gate software
sand-filled plastic bottle with string through the lid
metre rule
set square
force meter (to weigh bottle)
access to scales to find mass of bottle
G-clamp

The plastic bottle needs to be cylindrical and small enough to fit easily through the jaws of the light gate.

Filling it with sand makes it heavy enough to make air resistance unimportant on the first swing. A large black dot or ring halfway down the bottle marks the centre of mass of the bottle.

The table should provide a solid support for the top end of the pendulum.

Answers

Page 216

1. 100 J.
2. 4 J.
3. 812.5 J.
4. 8 kg.
5. 3 m/s.

Page 218 (top)

1. 100% efficiency.
2. Heat energy from friction.
3. 8.9 m/s.

Page 218 (bottom)

Student's own answers, but should include key points from the text.

P4c.2a Up and down

In this practical, you will investigate the link between transfer of energy and movement.

Apparatus

string

filled bottle

G-clamp

Method

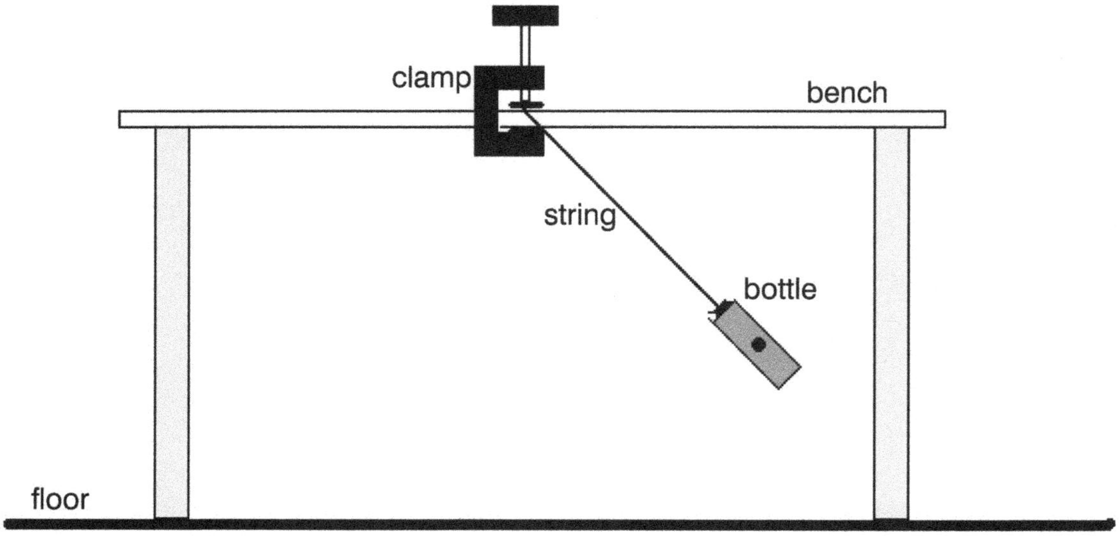

1. Clamp the end of the string to the centre of the bench so that the bottle swings just above the floor.
2. Pull the bottle to one side.

What type of energy are you transferring to it?

3. Release the bottle. Observe its behaviour.

Think about the energy transfers involved.

P4c.2b Pendulum swing

Apparatus

light gate
computer
string
metre rule
filled bottle
G-clamp
scales
force meter
set square

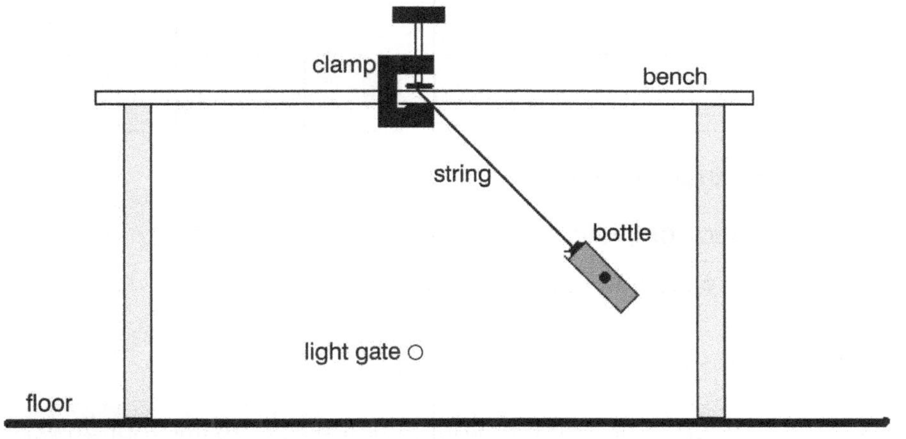

Method

1. Use the force meter to find the weight of the bottle and its contents. Don't try to remove the string.
2. Measure the mass of the bottle and contents.
3. Use the set square and ruler to measure the diameter of the bottle carefully (see diagram).

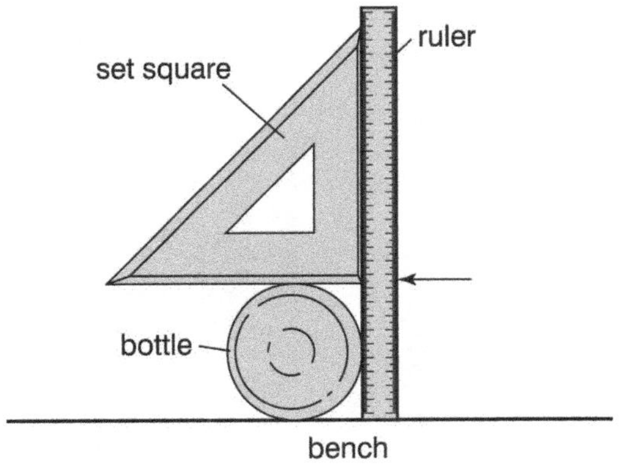

4. Clamp the end of the string to the centre of the bench so that the bottle swings just above the floor.
5. Arrange the light gate so that the bottle passes through it at the lowest point of its swing.
6. Set up the computer so that the light gate measures the speed of the bottle in m/s.
7. Mark the centre of the bottle with a dot.
8. Pull the bottle to one side as shown.

Measure the **initial height** of the dot above the floor.

9. Release the bottle so that it swings through the light gate. Record the speed of the bottle.
10. Measure the **final height** of the dot as it passes through the light gate.
11. Repeat steps 8 and 9 for a number of different initial heights. Try to cover a wide range of values from small to large.

12. Record your results in a table like this.

Initial height (m)	Final height (m)	Change of height (m)	Weight (N)	GPE lost (J)	Mass (kg)	Speed (m/s)	KE gained (J)

Analyse and interpret data

1. Do calculations to complete the table. Use these equations.

   GPE lost = weight × change of height

   $$\text{kinetic energy gained} = \frac{1}{2} \times \text{mass} \times \text{speed}^2$$

2. Do the results support or disprove your hypothesis? Justify your answer in detail.

Activity P4c.3 Power

Specification reference: P4c.4.16, P4c.4.17

Learning objectives

- Describe power as the rate of transfer of energy or the rate of doing work
- Use the relationship between power, work done (energy transferred) and time taken:

$$\text{power} = \frac{\text{work done}}{\text{time taken}}$$

$$P = \frac{W}{t}$$

Learning outcomes

- Be able to relate power to the rate at which energy is transferred

Common misconceptions

Like work, power has a precise meaning in science and a much looser meaning in everyday life. Not only does it have meanings that have nothing to do with energy transfer, it is often used to describe a force.

Resources

Student Book pages 218–223

Worksheet P4c.3a Personal power

Worksheet P4c.3b Power calculations

Resources for class practical (see Technician's notes, following)

Resources for initial demonstration: electric motor and mass.

Approach

Introduce the activity

Demonstrate the use of an electric motor to raise a mass up from the floor to bench height. Introduce the idea of power and make a quick measurement of its value for the motor.

Develop the activity

Discuss the concept of power as set out on pages 218–220 of the Student Book. Use the approach described above when considering the worked examples. Students need to be familiar with the idea of calculating work done followed by the power calculation if the subsequent practical work is to make much sense.

The practical on Worksheet P4c.3a can be used at this point. As usual, students can do this in small groups if equipment is available, or it can be done as a demonstration with active participation of the students. Be aware that some students can be very competitive when measuring their power.

It is important that they don't damage themselves by over-exertion. Some students will be unwilling to take part in physical activity – they could be allowed to borrow results from another student. After the practical, students can do the questions on Worksheet P4c.3b.

You may wish more able students to look at the Developing investigative skills feature on page 222 of the Student Book. The same safety considerations as mentioned above for the practical activity should be borne in mind for this activity.

Bring the activity together

Use class questioning to establish that students can demonstrate the true meaning of some key words: force, energy, work and power. Then discuss the questions on page 223 of the Student Book to check understanding.

Technician's notes

Be sure to check the latest safety notes on these resources before proceeding.

The following resources are needed for the class practical P4c.3a, per group:

variety of dumbbells
metre rule
access to bathroom scales (N or kg)
box or similar for students to step on to
variety of smaller masses
thread to tie mass to finger
stop clock

Do not use masses that are too large.

Answers

Worksheet P4c.3b

1. 10 W.
2. 120 000 J.
3. 5 s.
4. a) 7.1 m/s.
   b) 60 m.

Developing investigative skills, page 222

1

Lifting activity	Mass (kg)	Increase in height (m)	Energy transferred (J)	Time taken (s)	Power (W)
bag of sugar	1	2	20	0.6	33
bag of cement	25	0.5	12.5	1.6	78
bag of compost	7.5	0.8	60	0.6	100
watering can	15	1.6	20	3.2	75

2. a) Some results show that as height increases so does the power – e.g. when height increases from 0.5 to 0.8 m, power increases from 78 W to 100 W. Others show the opposite – e.g. when the height increases from 0.5 to 2 m, the power decreases from 78 W to 33 W..

   b) The results are too contradictory and incomplete to make certain conclusions.

3. a) Some results show that as mass increases, so does the power – e.g. from 1 to 25 kg, the power increases from 33 W to 78 W. Others show the opposite – e.g. from 7.5 to 25 kg the power goes down from 100 W to 78 W.

   b) The results are too contradictory and incomplete to make certain conclusions.

4. a) To investigate height use the same mass and lift it through different heights. Take the time for several lifts and find the average time for one lift.

   b) To investigate mass use the same height but lift different masses. Take the time for several lifts and find the average time for one lift.

Activity	**Mass (kg)**	**Height (m)**	**Energy (J) (mass × height × 10)**	**Time taken (s)**	**Power (W) (energy/time)**
lifting tin of food from floor to head height	0.4	2.4	9.6 J	1 s	9.6 W

Page 223

1. 240 W.

2. The man lifts twice the mass up the stairs, so he does twice the work. They take the same time, so the man must be providing twice the power.

3. It can transfer a lot of energy each second.

4. The watt.

5. a) 3000 J.

   b) 50 W.

P4c.3a Personal power

In this practical, you will investigate the power of your leg, arm and finger.

Apparatus

dumbbells

metre rule

stop clock

step

thread loop

masses

bathroom scales

SAFETY INFORMATION

Take care that masses are not dropped on feet.

Only lift masses you feel comfortable with. Do not over exert yourself.

Method

1. Find your **leg power** as follows:
   - Use the scales to find your weight in newtons.
   - Measure the height of the step in metres.
   - Calculate the work you do each time you stand on the step in joules.
   - Find out how many times you can stand on the step in a minute.
   - Divide the total work done by 60 to find your leg power in watts.
2. Find your **arm power** as follows.
   - Find a dumbbell which you can raise comfortably from your side to your shoulder using your best arm.
   - Use g = 10 N/kg to calculate its weight in newtons.
   - Measure the vertical distance you raise the dumbbell in metres.
   - Calculate the work you do each time you raise the dumbbell in joules.
   - Find out how many times you can raise the dumbbell in a minute.
   - Divide the total work done by 60 to find your arm power in watts.
3. Find your **finger power** as follows:
   - Place your hand so that the forefinger juts over the edge of the table.
   - Find a mass which you can raise comfortably with your forefinger.
   - Loop the thread around your finger so that it can raise and lower the mass.
   - Use g = 10 N/kg to calculate the weight of the mass in newtons.
   - Measure the vertical distance you raise the mass in metres.
   - Calculate the work you do each time you raise the mass in joules.
   - Find out how many times you can raise the mass in a minute.
   - Divide the total work done by 60 to find your finger power in watts.

P4c.3b Power calculations

1. Helen can lift a 50 N weight up a vertical distance of 0.30 m with her right arm. If she does this forty times in a minute, what is the power of her arm muscles?

2. A car engine has an output power of 6000 W. How much kinetic energy can the engine transfer to the car in 20 s?

3. A crane has a power of 2000 W. How long will it take to lift a 500 N block up by a distance of 20 m?

4. A car has a mass of 800 kg and engine power of 4 kW.

   a) It starts from rest. If the power of 4 kW is applied, what is the speed after 5 s? Assume there is no air resistance.

   b) The same car goes up a hill at a steady speed. If the power of 4 kW is applied, through what vertical height can the car climb in two minutes? g = 10 N/kg.

Activity P4c.4 Consolidation and summary

Learning objectives

- To review the learning points of the topic
- To test understanding through answering questions

Learning outcomes

- Be familiar with the knowledge and understanding summarised in the End of topic checklist
- Be able to apply this knowledge and understanding by answering End of topic questions

Resources

Student Book pages 224–226

Approach

Introduce the activity

Ask students to work with a partner to make a list of key words from this topic. They could then work together to produce a spider diagram showing how the different concepts are linked. They could compare their list with the list of key words given on page 224 of the Student Book. Discuss the checklist given on page 224 and use questioning to see how much of the content students are comfortable with.

Develop the activity

Ask students to work individually to work through the End of topic questions on pages 225–226 of the Student Book without reference to the text. As they work, walk around the classroom observing their answers and questioning them as necessary to find out which questions are causing difficulties.

Bring the activity together

After a set period, ask the students to stop working and discuss any areas of difficulty you observed as you walked round the class. Students should complete any unanswered questions for homework, but you should stress that they should attempt the questions without reference to the text to see how much content they have retained.

Answers

End of topic questions mark scheme

Question	Correct answer	Marks
1	12.5 m	1 1
2 a)	energy transferred = 90 J (per step) × 200 (number of steps) = 18 000 J	1 1
2 b)	power = energy transferred/time taken = 18 000/180 = 100 W	1 1
3 a)	10 500 J	1 1
3 b)	24.5 m/s	1 1
3 c)	some of the gravitational potential energy will have been transferred to thermal energy due to the friction between the sledge and the snow	1 1 1
4	row 1: 200 (J) row 2: 2 (N) row 3: 75 (m) row 4: 90 000 (J) row 5: 100 (N) row 6: 400 (m)	1 1 1 1 1 1
5	27 000 J	1 1
6	4 m	1 1
7	90 m	1 1
8	row 1: 3442.5 (J) row 2: 200 (kg) row 3: 8 (m/s) row 4: 80 275 (J) row 5: 3 (kg) row 6: 0.71 (m/s)	1 1 1 1 1 1
9	3.6 J	1 1

Question	Correct answer	Marks
10	0.77 m/s	1 1
11 a)	30 J	1 1
11 b)	1890 J	1 1
11 c)	1920 J	1 1
11 d)	at bottom of slope total KE = 1920 J KE = ½ mv^2 1920 = ½ × 60 × v^2 v^2 = 1920 ÷ 30 = 64 v = √64 = 8 m/s	1 1 1
12	This is a level of response question. For example: Level 1: When the speed doubles, the braking distance is four times	2
	OR when speed doubles, the braking distance more than doubles	1
	Level 2: When the speed doubles, the braking distance quadruples. There is more kinetic energy to be absorbed by the brakes when braking	4
	OR when the speed doubles, the braking distance is four times. Kinetic energy is reduced when braking	3
	Level 3: Kinetic energy = ½ mv^2 so at double the speed there is four times as much kinetic energy to be transferred on braking. The work done is equal to the decrease in kinetic energy. Work done = force × distance, so the braking distance quadruples	6
	OR when the speed doubles, the braking distance quadruples. At double speed, there is four times the amount of kinetic energy (½ mv^2) to be transferred on braking	5
	Total:	65

P4d Human influences on the environment

Introduction

This brief topic considers renewable and non-renewable resources for the generation of electricity.

Links to other topics

Section	Essential background knowledge	Useful links
2 Electricity		P2b Mains electricity
4 Energy resources and energy transfers	P4b Energy P4c Work and power	
6 Magnetism and electromagnetism		P6d Electromagnetic induction

Topic overview

P4d.1	**Energy resources** Students consider the various options for resources to generate electricity. They consider the advantages and disadvantages of methods of large-scale electricity production from various non-renewable and renewable resources.
P4d.2	**Consolidation and summary** This activity provides an opportunity for a quick recap on the ideas encountered in the topic as well as time for the students to answer the End of topic questions in the Student Book.

Activity P4d.1 Energy resources

Specification reference: P4d.4.18P, P4d.4.19P

Learning objectives

- Describe the energy transfers involved in generating electricity using: wind, water, geothermal resources, solar heating systems, solar cells, fossil fuels, nuclear power
- Describe the advantages and disadvantages of methods of large-scale electricity production from various renewable and non-renewable resources

Learning outcomes

- Understand the difference between energy sources and energy resources
- Be able to describe advantages and disadvantages of a range of energy resources, giving suitable examples

Common misconceptions

There is no such thing as an energy source which will last for ever, no matter how much energy you transfer out of it. However, in practice, energy sources that rely on the Sun or the Earth's core to replace energy transferred out of them will last at least a billion years – much longer than those that don't.

Resources

Student Book pages 227–234

Worksheet P4d.1a Investigating a model wind turbine

Resources for demonstration and class practical (see Technician's notes, following)

Approach

Introduce the activity

Ask students to explain why coal-fired power stations will become a thing of the past but those that are wood-fired will not.

Use a range of items (see list in Technician's notes) to explain the meaning of 'renewable' in the context of energy sources – track each one back to energy transferred from the Sun.

Develop the activity

Discuss 'Fossil fuels' on pages 227–228 of the Student Book.

Discuss why renewable resources are becoming increasingly important. Demonstrate a simple wind turbine made out of a motor and a propeller and show how its output depends upon how fast it spins. Put together a simple wind turbine using a motor and a propeller and, using a hairdryer, investigate how its output depends upon how fast it spins.

Students can then put together a simple wind turbine (made out of a motor and a propeller) and investigate the effect of speed on this simple wind turbine for themselves. Hairdryers used as sources of wind should have their heating circuits deactivated. Instructions are given on Worksheet P4d.1a.

Discuss how renewable resources are used in your country and why there may be opposition to some schemes (wind turbines and their aesthetic appearance, for example). More able students can then tackle the Developing investigative skills box about wind power on page 229 of the Student Book.

Now move on to discuss other renewable sources using pages 228–232 of the Student Book as a guide. More able students could look at the Science in context feature on the Severn Barrage on page 230 of the Student Book.

Bring the activity together

Discuss the questions on page 234 of the Student Book to check understanding. Also of use is the Science in context about future energy supplies which starts on page 232.

Technician's notes

Be sure to check the latest safety notes on these resources before proceeding.

The following resources are needed for the class demonstration to introduce the activity P4d.1:

calculator with solar panel	wooden splint
yeast fermenting sugar solution	green plant
spirit burner containing ethanol	

Note: Ethanol is HIGHLY FLAMMABLE and harmful if swallowed.

The following resources are needed for the demonstration in the main part of the activity:

hairdryer (with heating circuit deactivated)
small electric motor to act as dynamo
plastic propeller to fit onto motor shaft
voltmeter 0–50 mV (preferably moving coil)
leads and crocodile clips to connect motor to meter
clamp and stand to support motor

The following resources are needed for class practical P4d.1a, per group:

hairdryer (with heating circuit deactivated)	leads and crocodile clips to connect motor to meter
small electric motor to act as dynamo	clamp and stand to support motor
plastic propeller to fit onto motor shaft	anemometer
multimeter	metre rule

Answers

Developing investigative skills, page 229

1. Students' own ideas.

2. Students' own table.

3. Students' own ideas.

Page 230

Students' own report.

Page 234

1. Light energy to electrical energy and heat energy.

2. Kinetic energy in wind to kinetic energy in turbine to electrical energy and heat energy.

3. A fuel is used to turn water into steam. The steam has kinetic and thermal energy and drives a turbine. The turbine drives a generator which transfers the kinetic energy to electrical energy.

4. Using heat energy from the Earth.

P4d.1a Investigating a model wind turbine

You are going to plan an investigation to evaluate wind power as an energy source.

Apparatus

model wind turbine

multimeter to measure the voltage generated

anemometer to measure wind speed

hairdryer with heating circuit deactivated to generate wind power

metre rule to measure distance

SAFETY INFORMATION

Make sure that you show your plan to your teacher before you carry out the investigation. You should identify any risks and how you will control them in your plan.

Method

1. Plan your experiment, describing clearly the following:
   - the aim of your investigation
   - what you will measure
   - the number and range of readings that you will take
   - the independent variable
   - the dependent variable
   - the control variables
   - how you will make your experiment a fair test.
2. Draw out a results table that you would use in your investigation.

Analyse and interpret data

1. Look at your results. Is there any pattern in your readings?
2. Does your data support your hypothesis?

Evaluate data and methods

1. Write an evaluation identifying aspects of your experiment where modifications are possible.

Activity P4d.2 Consolidation and summary

Learning objectives

- To review the learning points of the topic
- To test understanding through answering questions

Learning outcomes

- Be familiar with the knowledge and understanding summarised in the End of topic checklist
- Be able to apply this knowledge and understanding by answering End of topic questions

Resources

Student Book pages 235–237

Approach

Introduce the activity

Ask students to work with a partner to make a list of key words from this topic. They could then work together to produce a spider diagram showing how the different concepts are linked. They could compare their list with the list of key words given on page 235 of the Student Book. Discuss the checklist given on page 235 and use questioning to see how much of the content students are comfortable with.

Develop the activity

Ask students to work individually to work through the End of topic questions on pages 236–237 of the Student Book without reference to the text. As they work, walk around the classroom observing their answers and questioning them as necessary to find out which questions are causing difficulties.

Bring the activity together

After a set period, ask the students to stop working and discuss any areas of difficulty you observed as you walked round the class. Students should complete any unanswered questions for homework, but you should stress that they should attempt the questions without reference to the text to see how much content they have retained.

Answers

End of topic questions mark scheme

Question	Correct answer	Marks
1 a)	a source that cannot be regenerated – it takes millions of years to form	1 1
1 b)	coal, oil and natural gas	1 1 1
1 c)	coal it is estimated that there is about 300 years' supply	1 1
2 a)	the total energy used over this period has increased there was a reduction in energy used from 1980 to 1983 but since then the consumption has increased	1 1 1
2 b)	the use of oil reduced up until about 1985 but has remained fairly constant since then the use of coal has reduced as the use of natural gas and nuclear power has increased	1 1 1
2 c)	I would expect energy consumption to continue to increase I would expect the use of renewable energy sources to increase the use of coal to decrease	1 1 1
3 a)	any two from: strength of the wind high ground constant supply of wind open ground	2
3 b)	advantage: renewable energy source, no air pollution disadvantage: unsightly, takes up too much space	1 1
4	**renewable resources** advantages: wind power and solar power are very efficient hydroelectric power can be used to store energy disadvantages: wind power only works when the wind is blowing solar power only works during the day **non-renewable resources** advantage: can be built close to population centres disadvantage: one of: fuel supplies will run out efficiency is only about 40%	 1 1 1 1 1 1
5 a)	the wind does not always blow sometimes it blows but not hard enough to generate electricity at other times it can be blowing too hard and the turbine has to be shut down	1 1 1
5 b)	no need to transport fuel to the area no need for expensive power cables to transmit the power to the area	1 1 1

Question	Correct answer	Marks
6 a)	coal oil natural gas	1 1 1
6 b)	advantage: close to population centres disadvantage: polluting	1 1
7 a)	coal	2
7 b)	coal	2
8	transport links size of site ease of link to national grid	1 1 1
9 a)	cost of running cost of building the power station	1 1
9 b)	oil increase hydroelectric stay same	1 1
9 c)	transferred to electrical energy some wasted as heat energy	1 1
9 d)	hydroelectric	2
9 e)	coal: advantage – can be sited near population, disadvantage – polluting hydroelectric: advantage – renewable, disadvantage – requires large area to be flooded for reservoir	1 1
10	students' own report	5
	Total:	59

Section 4: Answers to exam-style questions

Question	Correct answer	Marks
2 a) i)	current = 3 A voltage = 12 V	1 1
2 a) ii)	12 × 3 × 600 21 600 (J)	1 1
2 b) i)	efficiency = useful energy out ÷ total energy in	1
2 b) ii)	18 500 ÷ 21 600 0.86/86%	1 1
2 b) iii)	energy transferred to surroundings	1 1
2 c) i)	electrical heat/thermal heat/thermal surroundings	1 1 1 1
2 c) ii)	students' own Sankey diagram	3
3 a) i)	GPE = mass × *g* × height	1
3 a) ii)	30 × 10 × 3 900 (J)	1 1
3 a) iii)	equal	1
3 a) iv)	time taken	1
3 b)	KE gained = GPE lost equation for KE correct substitution correct value	1 1 1 1
4 a) i)	suitable scales plotting line if drawn/comment otherwise	2 2 1
4 a) ii)	reasonable comment, for example: lower values tend to be a little lower but not clear pattern OR pattern inconsistent so cannot state this conclusion	2
4 b)	student incorrect rate of cooling changes at different temperatures so starting difference will alter temperature change	1 1 1
4 c) i)	valid results allow you to answer the original question	1
4 c) ii)	with no lid convection can happen (easily) much more significant than conduction (through walls) so not testing the original question	1 1 1 1
	Total:	40

Contents

Overview of the section

In the first topic of this section students learn the relationships between density, mass and volume and pressure, force and area. They learn about experiments to determine density using direct measurements of mass and volume.

They discuss the fact that pressure in a point in a liquid or gas which is at rest acts equally in all directions. Finally they learn and use the relationship between pressure difference, height, density and '*g*'.

The second topic is one which considers changes of state. Students describe the arrangement and motion of particles in solids, liquids and gases. They also explore the qualitative and quantitative ideas of specific heat capacity and apply this to heating systems used in the home.

The final topic in this section discusses ideal gas molecules. This topic includes an introduction to the gas laws, as well as a definition of the kelvin temperature scale. Students learn why there is an absolute zero of temperature. They will learn that an increase in temperature results in an increase in the average speed of the molecules.

Starting points

The Student Book provides a 'section opener', a double-page spread which sets the scene for students (see pages 244–245). It is structured in this way:

- an overview providing details of the areas of study
- six questions, the first three relating to aspects of physics that should be reasonably familiar to most students, and the final three relating directly to some of the topics to be studied
- a list of section contents.

The questions provide a structure for introducing the section.

1. Describe how particles are arranged in (a) a solid, (b) a liquid and (c) a gas.
2. What happens to the particles when a solid melts?
3. Explain what happens when a solid dissolves in a liquid.
4. How are density, mass and volume related?
5. Are evaporation and boiling the same thing? Give a reason for your answer.
6. What happens to the speed of gas molecules as temperature increases?

The section opener has two main purposes:

- to acknowledge the student's prior learning and to value it
- to provide a benchmark against which future learning can be compared.

The six questions or 'starting points' in the Student Book can be used in a number of different ways to introduce the section:

- You could ask students to consider the questions as an introductory homework task.
- You could put students into groups to share their own ideas and understanding and then report back to the whole class.
- Students could be given access to the Internet, preferably with a tight timescale, to find out the information required.

You could then use a spider chart or other form of wall chart to summarise everybody's ideas. Recording these initial ideas in this way allows you to retain them for reference as the individual topics are developed. In this way, your students' progress in learning can be readily acknowledged.

P5b Density and pressure

Introduction

In this topic, students learn about density and pressure and the relationship between mass, volume and height. They will look at the differences between pressure in liquids and gases.

Links to other topics

Section	Essential background knowledge	Useful links
5 Solids, liquids and gases		P5c Change of state P5d Ideal gas molecules

Topic overview

P5b.1	**Density, mass and volume** In this activity students will explore the relationship between density, mass and volume.
P5b.2	**Pressure, force and area** In this activity students will explore the relationship between pressure, force and area.
P5b.3	**Pressure in gases and liquids** In this activity students will learn that pressure at a point in a liquid which is at rest acts equally in all directions. They will learn and use the relationship: pressure difference = height × density × g
P5b.4	**Consolidation and summary** This activity provides an opportunity for a quick recap on the ideas encountered in the topic as well as time for the students to answer the End of topic questions in the Student Book.

Activity P5b.1 Density, mass and volume

Specification reference: P5b.5.3, P5b.5.4

Learning objectives

- Know and use the relationship between density, mass and volume:

  $$\text{density} = \frac{\text{mass}}{\text{volume}}$$

  $$\rho = \frac{m}{V}$$

- Describe experiments to determine density using direct measurements of mass and volume

Learning outcomes

- Be able to calculate densities
- Be able to use ideas about density in a variety of situations, for example, to describe why convection currents occur

Resources

Student Book pages 247–253

Worksheet P5b.1a Finding the density of an irregularly shaped object

Resources for class practical (see Technician's notes, following)

Resources for class demonstration in introduction: metal and plastic teaspoons.

Approach

Introduce the activity

Show students a plastic teaspoon and a metal one (of approximately the same size). Ask students why they do not weigh the same (metal has higher density than plastic – if volume is the same, then mass of metal object will be greater than mass of plastic object).

Develop the activity

Discuss the content about density on pages 247–253 of the Student Book. Discuss how the formula for density can be used to determine the density of a regularly shaped object, and then give the students the opportunity to work through the worked example on page 248. After a short period, go through the students' solutions and compare them with the given solution and discuss where they differ and why.

Now consider the table of densities on page 249 and give students a series of questions about whether given materials will float or sink on given liquids. Students have to use the table to justify their answers. Students can work in pairs for this part of the activity, and stay in these pairs to tackle the questions on page 249 to check understanding of the activity so far.

Next, consider why materials have different densities. Ask students to work in pairs to explain why expanded polystyrene cups have a lower density than other cups. Allow students to tackle the worked example on page 251 in the way described above.

At this point, if you have the resources, students could measure the density of an irregular object in small groups, using the instructions given on pages 251–252 of the Student Book, or you could do this as a demonstration, as long as students are actively involved, for example, in weighing the object, taking readings for the volumes of liquid, and carrying out the calculations.

Instructions are also given on Worksheet P5b.1a.

As an alternative, students could work through the Developing investigative skills feature on page 253 of the Student Book. The worked example on page 252 should be tackled in the same way as previous examples in this topic, before the students tackle the questions on pages 252–253 to check understanding.

Bring the activity together

Ask students to work with a partner to make a summary of the key points of this activity. They could then compare their summary with another pair and make additions if necessary.

Technician's notes

Be sure to check the latest safety notes on these resources before proceeding.

The following resources are needed for the class practical P5b.1a, per group:

irregularly shaped object
spring balance
measuring cylinder
plastic and metal teaspoon
balance for demo

Answers

Page 249

1. a) 40 cm^3.
   b) 7.8 g/cm^3.
2. The bread contains more air spaces, making the overall density less.
3. It is lower than the density of the sea water.

Pages 252–253

1. When the object floats, absorbs the liquid (like a sponge) or is damaged by it (such as dissolving).
2. Narrow as possible to give greater movement of level, fine scale to help with precision.
3. 1.25 g/cm^3.
4. 78 g.
5. 7.14 cm^3.

Developing investigative skills, page 253

1. Measure the length of each side (in different places to check if shape really is regular) and then multiply length × width × height.

2. Checking the zero improves the accuracy – it makes it more likely that the mass value recorded is the correct value.

3. With the values given, the most sensible unit is g/cm^3, but could convert to metres and kilograms. The values are aluminium 2.7 g/cm^3, brass 8.4 g/cm^3, copper 8.9 g/cm^3, iron 7.9 g/cm^3.

4. Two significant figures is appropriate, since the volume is only given to a precision of two significant figures.

5. For an irregular object, the mass can still be found using a balance. However, the volume now needs a different method. Typically this will involve displacing water, either in a measuring cylinder or a eureka can, and finding the volume of the water displaced when the object is inserted. This method will not work if the object is soluble or absorbs water.

P5b.1a Finding the density of an irregularly shaped object

In this practical you will find the density of an irregularly shaped object.

Apparatus

balance

measuring cylinder wide and deep enough to hold object

irregularly shaped object

SAFETY INFORMATION
Water is a suitable liquid for this investigation.

Method

1. Find the mass, *m,* of the object using a balance as shown.

Record its mass.

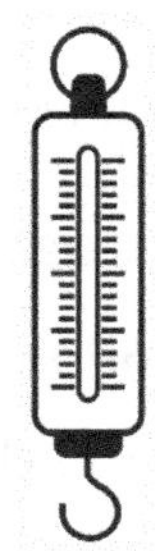

2. Choose a measuring cylinder that is wide and deep enough to hold the object.
3. Add liquid to fill the cylinder to a deep enough level so that the object will be completely submerged.
4. Measure the volume of liquid, V_1. Record it.
5. Lower the object into the liquid (without splashing).
6. Measure the new volume V_2. Record it.

Analyse and interpret data

1. Calculate the volume of the object $V_2 - V_1$. Record it.
2. Calculate the density of the object from the mass and the volume. Record it.
3. Why will a narrower cylinder give a more accurate answer than a wider one?
4. Why is the volume of the object $V_2 - V_1$?

Evaluate data and methods

1. How could you check your value for the density?
2. Why is it important that you do not splash when lowering the object into the liquid?

Activity P5b.2 Pressure, force and area

Specification reference: P5b.5.5

Learning objectives

- Know and use the relationship between pressure, force and area: pressure = $\frac{\text{force}}{\text{area}}$

$$p = \frac{F}{A}$$

Learning outcomes

- Be able to calculate pressures

Resources

Student Book pages 254–256

Resources for class demonstration (see Technician's notes, following)

Approach

Introduce the activity

Ask the students to look at the picture of the snowmobile (without reading the text around it) and ask them why it does not sink in the snow when, if its rider stood on the snow, he would probably sink into it, even though the snowmobile is much heavier than its rider. Draw out the fact that the snowmobile is covering a much larger area than its rider on the snow, and therefore the pressure it exerts on the snow is less.

Develop the activity

Ask the students to work in pairs to devise a way of demonstrating how pressure depends on force applied and area over which the force acts. For example, they might want to think about how they could model the situation of the snowmobile and the rider considered at the start of the activity.

In this case, what could they use to model the snow? How could they model the snowmobile? What about the rider? If you are able to provide materials such as those listed in the Technician's notes, then the students could carry out their demonstration if time allows.

Allow time for the students to tackle the worked example on page 255 of the Student Book – initially without the worked solution and then compare students' answers with the worked solution. Ask students to read the Developing investigative skills feature on pages 255–256 and answer the questions with a partner. After a short period, bring the class back together and discuss students' responses and compare them with the model responses. Discuss why the two sets differ (if they do).

Bring the activity together

Discuss the questions on page 255 of the Student Book to check understanding.

Technician's notes

Be sure to check the latest safety notes on these resources before proceeding.

The following resources are needed for the demonstration P5b.2:

material such as wood chips which can be used to model snow
a mass with a wide base to model the snowmobile
a mass with a lower value but narrower base to model the rider

Answers

Page 255

1. For the pin, the force is concentrated over a smaller area – there is a greater pressure.
2. 500 Pa.
3. 80 N.
4. 0.78 m^2.

Developing investigative skills, pages 255–256

1. Pressure = force ÷ area so different pressures can be achieved by changing either the force (by adding more 100 g masses), the area (by standing the block on different faces) or both.

2. Can be quite difficult to do reliably, but would need to have some way to level the surface (e.g. scrape with a rule) and some way to estimate depth – possibly by having a scale attached to the block. Would need to measure in several points around the block to find an average depth.

3. Independent variable is the pressure, dependent variable is the depth the block sinks.

4. Sand is a solid, even though it is made of quite small particles. As the block pushes down, there is little space between the grains of sand and they cannot be compressed. So if the downwards pressure on the block is high enough, the only thing the sand can do is move out to the side, making it more difficult to find the average depth of sinking.

5. Need a way to measure how much water has been added. Particularly helpful to have a container with no drainage holes at the bottom, then you know that all the water is in there somewhere. Use a measuring cylinder to add a known volume of water to the sand and carry out the test.

6. A camel's foot is an example of a load being spread over a wider area to reduce the pressure.

Activity P5b.3 Pressure in gases and liquids

Specification reference: P5b.5.6, P5b.5.7

Learning objectives

- Understand how the pressure at a point in a gas or liquid which is at rest acts equally in all directions
- Know and use the relationship for pressure difference:

$$\text{pressure difference} = \text{height} \times \text{density} \times g$$
$$p = h \times \rho \times g$$

Learning outcomes

- Be able to use ideas about pressure in a variety of situations, for example, to explain why pressure increases as you go towards the bottom of the ocean
- Be able to calculate pressure differences

Resources

Student Book pages 256–261

Resources for class demonstration in activity introduction: bottle of water

Approach

Introduce the activity

Show students a bottle containing water. Ask whether the pressure at the top of the bottle is the same as that at the bottom of the bottle. If not, which is greater? (No, the pressure is greater at the bottom.)

Develop the activity

Discuss the content on pages 256–257 of the Student Book. Discuss the questions on page 258 to check understanding.

Remind students about the question you asked about the pressure in the bottle of water at the beginning of the activity. Discuss the content on pages 258–260 of the Student Book. Again, students should tackle the worked example without the given solution and then compare their response with the model version. Students should discuss the questions on page 261 with a partner.

Bring the activity together

Ask students to work with a partner to make a summary of the main points of this activity. You may wish them to compare their summaries with another pair, or you may bring the class together to discuss the points that the students have noted. Question them to find out why they have put down the points they have. Justifying their answers is an important skill for them to develop.

Answers

Page 258

1. Particles colliding with other particles and the walls of the container.
2. The same air pressure is in our lungs and presses outwards.
3. The pressure outside the bottle is greater than the pressure inside the bottle.

Page 261

1. Depth, density of the fluid, gravitational field strength.
2. 80 000 Pa (80 kPa).
3. 1 030 000 Pa (1030 kPa).
4. 400 kPa.
5. 20 400 Pa (20.4 kPa).

Activity P5b.4 Consolidation and summary

Learning objectives

- To review the learning points of the topic
- To test understanding through answering questions

Learning outcomes

- Be familiar with the knowledge and understanding summarised in the End of topic checklist
- Be able to apply this knowledge and understanding by answering End of topic questions

Resources

Student Book pages 262–264

Approach

Introduce the activity

Ask students to work with a partner to make a list of key words from this topic. They could then work together to produce a spider diagram showing how the different concepts are linked. They could compare their list with the list of key words given on page 262 of the Student Book. Discuss the checklist given on page 262 and use questioning to see how much of the content students are comfortable with.

Develop the activity

Ask students to work individually to work through the End of topic questions on pages 263–264 of the Student Book without reference to the text. As they work, walk around the classroom observing their answers and questioning them as necessary to find out which questions are causing difficulties.

Bring the activity together

After a set period, ask the students to stop working and discuss any areas of difficulty you observed as you walked round the class. Students should complete any unanswered questions for homework, but you should stress that they should attempt the questions without reference to the text to see how much content they have retained.

Answers

End of topic questions mark scheme

Question	Correct answer	Marks
1 a)	depends on the sample	1
1 b)	float	1
1 c)	depends on the sample	1
1 d)	float	1
1 e)	sink	1
1 f)	sink	1
1 g)	float	1
2	1900 kg/m^3	1
3	if the crown is pure gold, the new water level will be 900 cm^3 if the jeweller has cheated, the water level will be higher than 900 cm^3	1 1 1 1
4 a) i)	160 kPa	1 1
4 a) ii)	159 kPa	1 1
4 a) iii)	8000 kPa	1 1
4 b)	the high-heeled shoe will damage the floor	1
5 a)	he is 15 m deep	1 1
5 b)	the submarine would be slightly less than 15 m deep	1
6 a)	29.4 kPa	1
6 b)	189.4 kPa	1
7	7.2×10^5 Pa	1 1
8	1.44×10^7 Pa	1 1
9	$(2.96 \times 10^5 - 1.00 \times 10^5) \div (1000 \times 10)$ = 19.6 m	1 1 1
10	7.2×10^{-3} kg	1 1

Question	Correct answer	Marks
11 a)	$p = \rho gh$ (pressure = density × acceleration due to gravity × depth)	2
11 b)	5000 Pa	1 1
12	1750 m	1 1
13	2.23×10^5 Pa	1 1
14	1.07×10^5 Pa	1 1
15	Larger weight of adult is concentrated on same small area Adult experiences higher pressure (and hence pain) Pressure on foot = $F \div A$ Pressure on child's foot = 120 ÷ 2 = 60 N/cm^2 Pressure on adult's foot = 600 ÷ 2 = 300 N/cm^2	1 1 1 1 1
	Total:	48

P5c Change of state

Introduction

In this topic students will learn about the changes that take place at a molecular level when a substance changes state.

Links to other topics

Section	Essential background knowledge	Useful links
5 Solids, liquids and gases	P5b Density and pressure	P5d Ideal gas molecules

Topic overview

P5c.1	**Arrangement of particles in solids, liquids and gases** In this activity students will learn about the arrangement and motion of particles in solids, liquids and gases.
P5c.2	**Melting and evaporation** In this activity students will learn about melting, evaporation, boiling and condensation, and the changes that occur during these processes.
P5c.3	**Specific heat capacity** This activity will use experimental data to determine the specific heat capacity (shc) of different liquids. It will also explore the practical difficulties in determining these values given the difficulties in reducing energy wastage. Also the discussions arising will develop the understanding and relevance of shc to the features of heating systems. The worksheet of calculations will also reinforce the idea of shc and the calculations associated with it.
P5c.4	**Consolidation and summary** This activity provides an opportunity for a quick recap on the ideas encountered in the topic as well as time for the students to answer the End of topic questions in the Student Book.

Activity P5c.1 Arrangement of particles in solids, liquids and gases

Specification reference: P5c.5.10P

Learning objectives

- Describe the arrangement and motion of particles in solids, liquids and gases

Learning outcomes

- Know how particles are arranged in solids, liquids and gases

Common misconceptions

Students often find it difficult to accept a model of a gas which leaves most of the space it occupies empty. They intuitively prefer a model in which the particles expand to fill the space available instead of a model in which they are free to move apart from one another to fit the space available.

Also, students often assume that particles in liquids and gases vibrate instead of moving around in random directions.

Resources

Student Book pages 266–267

Worksheet P5c.1a Kinetic theory

Resources for the class demonstration (see Technician's notes, following)

Resources for final class demonstration: flashcards with drawings representing the arrangement of particles in solids, liquids and gases.

Approach

Introduce the activity

Ask students to name the three states of matter. Issue students with a sealable sandwich bag and about six small polystyrene balls to place inside. Tell them to close the bag, squeezing all the air out, so that the balls are jammed together and cannot move, modelling a solid.

Letting a little air in should allow the balls to stay close but slide past each other, modelling a liquid. Then get them to seal the bag again, this time allowing lots of air into the bag, and to shake the bag around so that the balls model a gas.

Develop the activity

Discuss the content of pages 266–267 of the Student Book. Take every opportunity at this early stage to ensure students have a clear mental image of particles in the three states of matter, taking particular care to ensure that their picture of a gas is correct (not simply empty space; particles do not expand).

Get the whole class to act out the effect of heating a solid until it becomes a gas. Use this role play to develop the kinetic theory model, with an emphasis on how the arrangement of particles can account for the different large-scale properties of matter.

Discuss the idea of using bond strengths to explain the existence of all three different states of matter at room temperature and to explain changes of state. At this point, students can tackle the questions on the Worksheet, either individually or in pairs. It can be useful for students to work with a partner to discuss ideas.

Students need to realise that kinetic theory is a scientific model. It therefore has to be able to account for **all** observed properties of matter, not just some. Their answers to the questions need to clearly distinguish between an observation and its explanation with the model.

Bring the activity together

Discuss the questions on page 267 of the Student Book. Show students flash cards of various drawings representing the arrangement of particles in solids, liquids and gases. Ask them to say which state of matter each card represents and to explain how the state can be changed.

Technician's notes

Be sure to check the latest safety notes on these resources before proceeding.

The following resources are needed for the class demonstration P5c.1, per group:

sealable plastic sandwich bags
6 small polystyrene balls

The following resources are needed for the final class demonstration:

flashcards with drawings representing the arrangement of particles in solids, liquids and gases

Answers

Worksheet P5c.1a

1. The solid has circles clumped together in a regular pattern and touching; liquid has circles close to each other in a random pattern and not necessarily touching; gas has circles scattered at random throughout.

2. a) Each particle is held in place by the particles around it, so no change of position is possible.
   b) Each particle can move somewhere else, provided that it stays close to other particles, allowing change of shape without any overall change of volume.
   c) Particles are free to move anywhere in the container and do not have to stay near other particles.

3. a) Bonds between particles pull them towards each other. Energy is needed to pull them apart again. So give enough energy to particles in a solid and they will be able to get far enough apart to slide past each other – and become a liquid. Similarly, give enough energy to particles in a liquid and they will be able to escape each other completely and become a gas.
   b) Students' own definitions.
   c) i) Incorrect. There are bonds between particles in a gas.
   ii) Incorrect. The number of particles stays the same when a liquid becomes a gas.
   iii) Correct.
   iv) Incorrect. Particles in liquids move faster when their temperature is increased.
   v) Correct.
   vi) Incorrect. When a gas condenses it releases energy to bond the particles together.

Page 267

1. Increases – either as a larger vibration in solids or as faster translational motion in liquids and gases.

2. Compressing a gas pushes the particles closer together, in a liquid they are already close to each other and will repel if pushed closer.

3. a) Regular pattern, closely packed together, particles held in place.

   b) Irregular, closely packed together, particles able to move past each other.

   c) Irregular, widely spaced, particles able to move freely.

4. The size of the container it is put in.

P5c.1a Kinetic theory

1. Models of matter

a) Here are three boxes. One contains a solid, one a liquid and one a gas. Copy the three boxes and draw six small circles in each box to show the state of its particles.

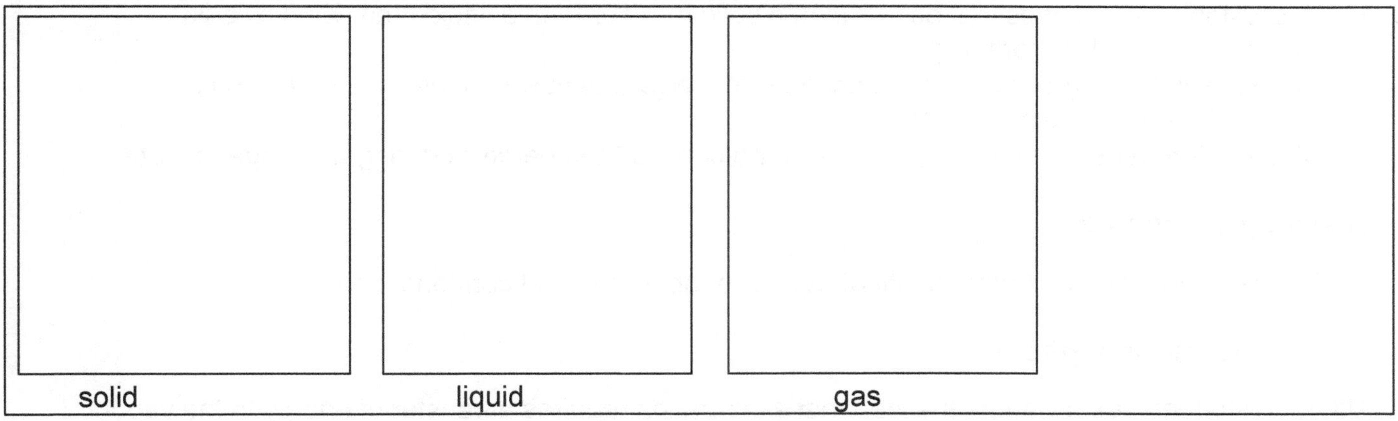

2. Explaining solids, liquids and gases

Use kinetic theory to write an explanation of why

a) solids have a fixed shape and size;

b) liquids have a fixed size but no fixed shape;

c) gases completely fill their container, whatever its shape or size.

3. Bonds and state changes

a) Use the idea of bonds to explain why energy has to be added to make a solid into a liquid or a liquid into a gas.

b) Define each of the following terms:

boiling, heating, melting, cooling, solidifying, condensing

c) Here are some statements about kinetic theory.

Decide if each one is correct or incorrect. For each incorrect statement, write a correct version.

i) There are no bonds at all between particles in a gas.

ii) The number of particles increases when a liquid becomes a gas.

iii) Solids expand when heated because their particles vibrate more.

iv) Particles in liquids vibrate more when their temperature is increased.

v) The bonds between particles of a gas are weaker than those of a liquid.

vi) When a gas condenses it takes in energy to bond the particles together.

Activity P5c.2 Melting and evaporation

Specification reference: P5c.5.8P, P5c.5.9P, P5c.5.11P

Learning objectives

- Understand the changes that occur when a solid melts to form a liquid, and when a liquid evaporates or boils to form a gas
- Explain why heating a system will change the energy stored within the system, raising its temperature or changing its state
- Obtain a temperature–time graph to show the constant temperature during a change of state

Learning outcomes

- Be able to explain the terms: melting, boiling, evaporation and condensation

Common misconceptions

Although students readily adopt the idea that evaporation requires a transfer of energy to the vapour from somewhere else, they are often reluctant to accept that condensation requires a transfer away from the vapour.

Many students think that the bubbles in a boiling liquid contain air. You might want to point out that bubbles of air do form below the boiling point as dissolved gases come out of solution.

Students are happy to accept the idea that when thermal energy is added to cold water its temperature will rise. Often though when its temperature reaches 100 °C students do not see an obvious use for the thermal energy that continues to be absorbed as the water boils. Relating this idea of latent heat to the behaviour of particles can help to explain how the energy is being used, despite there being no temperature increase. Similar misconceptions are common with melting too and time spent using kinetic theory will help acceptance here.

The cooling wax practical on Worksheet P5c.2a introduces more potential misconceptions. Many students will think that when things cool and freeze the temperature will simply fall as with a simple cooling curve. They will not expect the temperature to stabilise during freezing.

Freezing itself is normally associated with 'icy cold' temperatures and it is often assumed that all things freeze at temperatures near the freezing point of water. The connection between freezing temperature and melting point perhaps needs to be made. Then the freezing point of iron, for example (1538 °C), could be mentioned to further deal with the 'freezing equals icy cold' idea.

Resources

Student Book pages 267–271

Graph paper

Worksheet P5c.2a Candle wax cooling curve

Worksheet P5c.2b Evaporation and condensation

Resources for class demonstration (see Technician's notes, following)

Resources for class practical

Approach

Introduce the activity

Discuss students' experiences of melting and boiling, for example, melting butter or boiling a kettle. Half fill a large glass beaker with expanded polystyrene balls. Put it on the bench, grasp with both hands and shake it vigorously from side to side to simulate a liquid. Increase the agitation until some of the balls get enough energy to escape the beaker.

Discuss how this models evaporation.

The students will perform their own practical experiment to collect data for stearic acid as it cools, freezes and then continues to cool. Before this they will predict how the stearic acid will cool and sketch a cooling curve to describe their prediction. They will analyse, describe and explain the data they collect and relate this to the melting point.

Develop the activity

If possible, demonstrate the experiment as described on page 267 of the Student Book. Involve students with recording the readings and plotting the graph. Discuss the shape of the graph, using page 268 of the Student Book as guidance. Students can tackle the questions on page 269 in pairs.

Discuss the Developing investigative skills material on pages 269–270 of the Student Book. Look at the graph showing the experimental results and discuss the questions. Worksheet P5c.2a provides students with an opportunity to carry out this experiment.

Start by showing the students a lit candle. You can ask them about the flame and what is happening to the solid wax. They can suggest why it melts and freezes. You can point out that the temperature of the flame is anything between 800 °C up to 1400 °C. Its average temperature will be about 1000 °C. The temperature of the room will, however, only be about 20 °C. You can ask them to guess the freezing temperature of the wax and suggest how this could be measured practically. They can also use their ideas to predict and sketch a cooling curve for wax showing the liquid and solid phases. Using the practical ideas from the class and the predicted cooling curves they have produced, the basis of the students' experiment can be developed.

Candle wax is a mixture of substances so this experiment is better done with one of those substances – stearic acid (systematic name is octadecanoic acid, $C_{17}H_{35}COOH$). Further details are provided on Worksheet P5c.2a.

Now move on to discuss evaporation, using page 270 of the Student Book for guidance. Students can now tackle the questions on Worksheet P5c.2b.

You may wish more able students to look at the extension feature on page 271.

Bring the activity together

The students can plot the data from their experiment onto a graph to produce a curve. From this they should be able to identify the freezing point and notice the stable temperature at this point.

The class results can be discussed and the findings shared. This is an ideal opportunity to develop a kinetic explanation of the findings. The kinetic models can be used to good effect in showing what happens when:

- solids heat up – particles vibrate more energetically
- solids melt – particles gain enough KE to overcome the attractive forces between them and so they move apart and start to flow; this melting requires the addition of a considerable amount of energy
- liquids heat up – when all solid has melted the free moving particles move around faster.

The students can then be asked to use this model to explain the reverse, i.e. their own experiment when hot molten stearic acid cools, freezes to become solid and then cools further towards room temperature.

Students can then work with a partner to discuss the questions on page 271 of the Student Book and then work with the same partner (or a different one if appropriate) to construct a spider diagram to summarise their learning.

Technician's notes

Be sure to check the latest safety notes on these resources before proceeding.

The following resources are needed for the class demonstration P5c.2:

large glass beaker
expanded polystyrene balls
thermally insulated beaker
thermometer
heater
heating circuit (ammeter, voltmeter, battery, switch)
candle and matches

The following resources are needed for the practical P5c.2a:

eye protection
thermometers (0–100 °C) or temperature sensors and data logger
stearic acid (see advice from CLEAPSS)
stopwatches
test tubes
beakers (at least 400 cm^3)
clamp stand, boss and clamp
heatproof mats
graph paper
tongs or heatproof gloves

Answers

Worksheet P5c.2b

1. a) Evaporation and boiling transfer energy from the surroundings and involve liquid going to gas; condensing transfers energy to the surroundings and is a change from gas to liquid.

   b) i) Incorrect. When a liquid boils it forms bubbles of steam.

   ii) Incorrect. A liquid can evaporate at any temperature.

   iii) Correct. iv) Correct. v) Correct.

   c) The order is: iv), v), ii), iii), vi), i) vii).

2. a) Raising the temperature increases the average energy of the particles, making it more likely that a particle will get enough energy from a collision to break free from the bonds holding it to the rest of the liquid.

   b) Particles can only leave the liquid at its surface, so increasing the surface area increases the chances of them escaping from the liquid.

c) The air above the liquid can only hold a certain amount of vapour. Once that level has been reached, particles will re-enter the liquid as fast as they leave it. Continually moving the air away from the surface avoids this.

Page 269

1. Energy is needed to break the bonds between the particles in the ice.

2. All the particles have the energy to escape. In evaporation, only a fraction has.

Developing investigative skills, page 269

1. Candle wax is flammable.

2. 81 °C

3. At the start, the temperature of the wax is much higher than the room temperature. Rate of cooling decreases as the temperature difference reduces.

4. The wax is freezing or solidifying. The particles are moving slower so the forces are able to hold the particles in fixed positions.

5. a) Use a temperature sensor or probe. This could be linked to a data logger.

   b) The temperature measurements will be to the nearest 0.1 °C, although they may fluctuate due to measurement uncertainty. The estimate of melting point should be made to 0.1 °C. The method is also easier as it is automated.

6. The wax would cool slower as it has a greater mass and greater energy.

Page 271 (top)

1. Higher temperature, flow of air across the surface.

2. Larger area for the water molecules to evaporate from.

Page 271 (bottom)

Students' own answers, but should be along the following lines:

1. I changed from liquid state to gaseous state.

2. In the liquid state I was reasonably close to my neighbours, but not so close that I was right beside them – they were there if I needed them but we didn't live in each other's pockets! I was given some extra energy. The extra energy meant that I moved more quickly and was able to break completely away from my neighbours.

3. I escaped to the surface of the liquid and became a gas particle.

P5c.2a Candle wax cooling curve

This experiment involves water and stearic acid (a component of candle wax) at 80 °C. Care needs to be taken, as the wax will be very hot, and eye protection needs to be worn.

This experiment starts with the stearic acid as a liquid.

1. Using tongs or heatproof gloves, carefully collect a test tube of melted stearic acid that contains a thermometer.
2. Clamp the test tube or place it in a test tube rack. Make sure it is stable and cannot fall over.
3. Record the temperature every minute as the stearic acid cools. Continue taking measurements until the stearic acid reaches room temperature. This could take 30 minutes or so.
4. Plot a graph of temperature against time as the stearic acid cools.
5. Comment on the state of the stearic acid (liquid or solid) and annotate your graph.

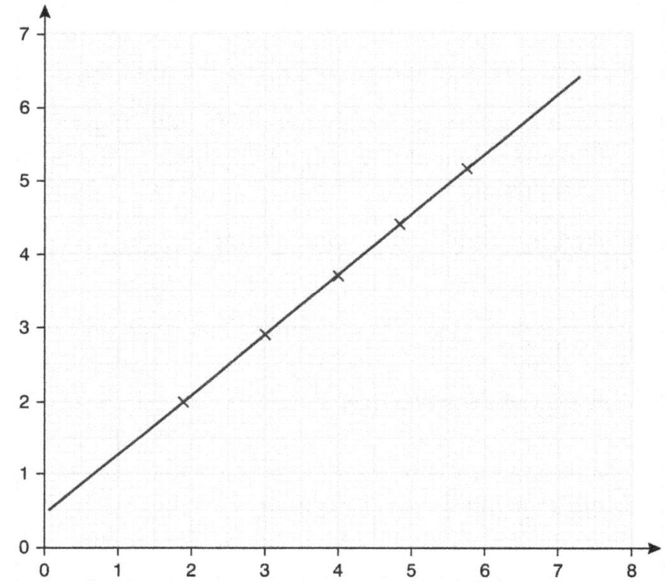

P5c.2b Evaporation and condensation

1. Evaporate, condense, boil

a) Draw lines to link each change of state to its description and energy transfer.

energy transfer	change of state	description
to surroundings	evaporation	liquid going to gas
	boiling	
from surroundings	condensing	gas going to liquid

b) Here are some statements about evaporation and condensation. Decide if they are correct or incorrect.

For each incorrect statement, write a correct version.

i) When a liquid boils, it forms bubbles of air.

ii) A liquid can only evaporate at its boiling point.

iii) A gas can only condense below the boiling point.

iv) Evaporation takes place only at the surface of a liquid.

v) A condensing gas transfers energy to its environment.

c) The sentences below explain how evaporation takes place. Put them in the best order.

i) The particle leaves the liquid.

ii) Each collision changes its energy.

iii) One collision leaves it with lots of energy.

iv) A particle is close to the surface of the liquid.

v) It collides many times with particles around it.

vi) It breaks free from the bonds with other particles.

vii) It becomes part of the gas above the surface.

2. Speed of evaporation

Use ideas about particles, bonds and energy to explain how each of the following increases the rate of evaporation of a liquid.

a) Increasing the temperature.

b) Increasing surface area.

c) Removing the air around it and replacing it with fresh air.

Activity P5c.3 Specific heat capacity

Specification reference: P5c.5.12P, P5c.5.13P, P5c.5.14P

Learning objectives

- Know that specific heat capacity is the energy required to change the temperature of an object by one degree Celsius per kilogram of mass (J/kg °C)
- Use the equation: change in thermal energy = mass × specific heat capacity × change in temperature
  $\Delta Q = m \times c \times \Delta T$
- Investigate the specific heat capacity of some materials

Learning outcomes

- Be able to determine experimental values of shc and compare these to accepted values
- Be able to explain the practical reasons that cause the differences between experimental values and accepted values

Common misconceptions

Many students are unaware that different materials take longer to heat up than others. They are therefore also unaware that different materials can store different amounts of energy. Confusions over heat and temperature can also complicate things. Added to this are the units for shc as J/kg °C. The experiment will at least help construct some understanding around these fragile ideas.

Resources

Student Book pages 271–274

Worksheet P5c.3a Specific heat capacity calculations

Resources for class demonstration or class experiment (see Technician's notes, following)

Approach

Introduce the activity

Standard electrical immersion heaters (usually 50 W at 12 V) can be used to heat say 100 g of water and 100 g of cooking oil. A suitable container such as a narrow beaker or small tin can (with sharp edges removed beforehand) can be used for the liquids. A joulemeter can be used to measure the energy output of the heater. If joulemeters are not available then students can measure the time in seconds. Given that the power is, say, 50 W (= 50 J per second) then the energy output can be calculated if the time in seconds is measured. Students can find out how much energy (equal to how many seconds × 50 W) is transferred when the liquid raises its temperature by say 10 °C. The calculations can be built up so as to illustrate what specific heat capacity is. For instance:

'How much energy is used to heat 100 g by 10 °C?'

'How much energy is used to heat 100 g by 1 °C?' (i.e. one-tenth of the amount)

'How much energy is used to heat 500 g by 10 °C?' (i.e. five times more)

The specific heat capacity for water is 4200 J/kg °C, vegetable oil is about 1670 J/kg °C and olive oil is 1970 J/kg °C.

Students can calculate their experimental values for the shc of water and oil and compare it to the accepted values. A useful discussion can follow which tries to explain why more energy from the heater was needed than expected. Answers would include: thermal energy escapes into atmosphere, is used to heat heater or the container. This discussion can lead to an improved plan for a repeat of the experiment. This can be run as a competition, with the winners the ones who get closest to the real value (without cheating!).

Develop the activity

Pages 271–272 of the Student Book usefully explore the idea of shc as applied to water-filled and oil-filled radiators in the home. The experiment that has just been done will help reinforce the ideas here. Water has a high specific heat capacity and therefore can require lots of energy to heat it by a few degrees. Oil, on the other hand, will require about half the energy per kg °C. So although, say, 10 kg of water will need twice the energy that 10 kg of oil needs to heat them by 40 °C, if it is in a radiator the water will then have more energy to give to the room. Because of its high shc water is therefore a good way to store lots of energy for space heating.

The students will need to understand and apply the relationship $\Delta Q = m \times c \times \Delta T$. They have already done this, to some extent, with the oil and water practical. However, it is worth showing students how to use the calculation to calculate energy. After this, manipulation of the formula can be introduced. There is a worked example on this in the Student Book, although further examples are always worthwhile. Further work to practise and reinforce these ideas can be found on Worksheet P5c.3a.

Bring the activity together

The 'Developing investigative skills' material on pages 273–274 of the Student Book shows an experiment to find the shc of some metals. This can be discussed and attempted by the students. Useful parallels can be drawn between this and the oil and water experiment.

Technician's notes

Be sure to check the latest safety notes on these resources before proceeding. This is especially important given that power supplies and a (small amount of) water are being used at the same time.

The following resources are needed for the class demonstration P5c.3:

12 V electric immersion heaters, leads and suitable power supply
thermometers or temperature sensors
cooking oil and water
joulemeters or stopwatches
narrow beakers or small tin cans
balance to measure 100 g (± 1 g)

Answers

Worksheet P5c.3a

1. a) D [1]

   b) D [1]

   c) Any two from:

Most heat needed to raise temperature / AW. [1]

least temperature rise per energy unit / AW. [1]

lowest gradient. [1]

2. a) (Highest) Water. [1]

(Lowest) Lead. [1]

b) 1800 [1]

770 [1]

8000 [1]

2 [1]

2 [1]

Aluminium [1] 900 [1]

3. $\Delta Q = m \times c \times \Delta T = 5 \times 900 \times 80 = 360\ 000$ J.

4. $\Delta Q = m \times c \times \Delta T$

50 kJ is 5000 J

$$c = \frac{\Delta Q}{m \times \Delta T}$$

= 50 000 = 444.4 J/kg °C (or 444 J/kg °C).

5. 1.5 × 75

$\Delta Q = m \times c \times \Delta T$

A 7 kW heater transfers 7000 J of energy each second. In 100 minutes it will transfer 7000 × 100 × 60 = 42 000 000 J.

$$\Delta T = \Delta T = \frac{\Delta Q}{m \times s} = \frac{42\,000\,000}{300 \times 4200} = 33.3\,°\text{C}$$

Temperature rise = 33 °C.

Final average temperature = 16 °C + 33 °C = 49 °C.

6. ΔQ (water) = $m \times c \times \Delta T = 2 \times 4\ 200 \times 80 = 672\ 000$ J

ΔQ (aluminium) = $m \times c \times \Delta T = 2 \times 900 \times 80 = 144\ 000$ J

Total energy = ΔQ (water) + ΔQ (aluminium) = 672 000 +144 000 = 816 000 J.

Developing investigative skills, page 274

1. aluminium = 11 500 J

copper = 4500 J

lead = 1500 J

steel = 6000 J

2. To give good contact between the thermometer and the metal. More thermal energy would escape if air was used.

3. Mass of the block, temperature rise (10 °C) and the power of the heater (50 W).

4. aluminium = 1150 J/kg °C

copper = 450 J/kg °C

lead = 150 J/kg °C

steel = 600 J/kg °C

5. All his experimental values are higher than the researched ones. More energy than expected appears to have been used to heat the blocks. This extra energy was probably used to heat the air and surroundings.

6. Insulate the blocks so less energy is transferred through the table and into the air.

7. a) Measure a fixed mass of cold water (e.g. 1 kg) into a large beaker.

b) Measure and record the initial temperature of the water.

c) Heat the water with an immersion heater for a fixed period of time, e.g. 10 minutes.

d) Stir, so that the water is at a uniform temperature throughout.

e) Measure and record the temperature at the end.

f) Work out the change in temperature.

g) Repeat for several other masses of water.

h) Measure the energy transferred by the heater to the water using a joulemeter, or calculate this from the power of the heater and the time it was switched on for.

i) Use the equation $\Delta Q = m \times c \times \Delta T$ to calculate *c,* the specific heat capacity of water.

P5c.3a Specific heat capacity calculations

1. Jane investigates specific heat capacity of four materials: A, B, C and D.

She heats the same amount of different materials and measures their temperatures.

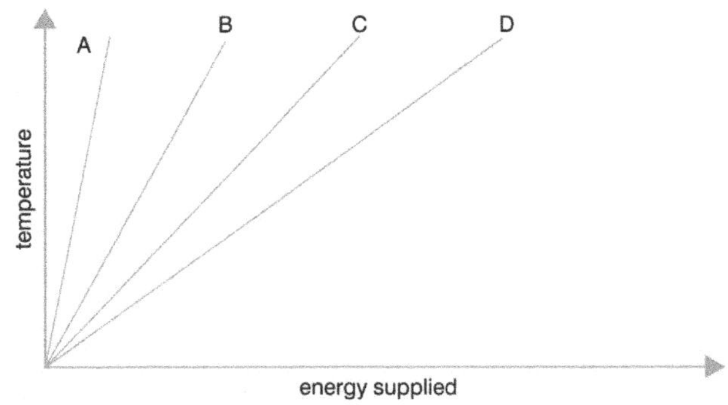

a) Which material needs most energy to raise its temperature by 1 °C?

Circle your answer.

A B C D

[1]

b) Which material needs least energy to raise its temperature by 1 °C?

Circle your answer.

A B C D

[1]

c) Material D has the highest specific heat capacity. Explain how you can tell this from the graph.

[2]

2. a) Look at the table. It shows the specific heat capacities (shc) of different materials.

Material	**Specific heat capacity in J/kg °C**
aluminium	900
copper	385
lead	160
steel	450
water	4200

Complete the sentences below.

The material with the **highest** SHC is . . . .

The material with the **lowest** SHC is . . . . [2]

b) You can use this equation to calculate the energy supplied to different materials.

energy = mass × specific heat capacity × temperature rise

Use this information and the equation to complete the table below.

Material	**Mass (kg)**	**Specific heat capacity (J/kg °C)**	**Temperature rise (°C)**	**Energy needed (J)**
aluminium	2	900	1	
copper	1	385	2	
lead	10	160	5	
steel		450	10	9000
water	5	4200		42 000
	5		10	45 000

3. Aluminium has a specific heat capacity of 900J/kg °C. Calculate the minimum amount of energy needed to heat a 5 kg piece of aluminium from 20 °C to 100 °C.

4. A 1.5 kg iron pan absorbs 50 kJ from a cooker hob and increases it temperature by 75 °C. Calculate the specific heat capacity of the iron.

5. A hot water storage tank contains 300 kg of water at 16 °C. A 7 kW heater takes 100 minutes to transfer energy to heat the water. Calculate the maximum average temperature reached by the water if the specific heat capacity of water is 4200J/kg °C.

6. A 2 kg pan is made of aluminium and contains 2 kg of water. The pan and its water are heated from 20 °C to 100 °C. Calculate the energy needed for this if the specific heat capacity of water is 4200J/kg °C and aluminium is 900 J/kg °C.

Activity P5c.4 Consolidation and summary

Learning objectives

- To review the learning points of the topic
- To test understanding through answering questions

Learning outcomes

- Be familiar with the knowledge and understanding summarised in the End of topic checklist
- Be able to apply this knowledge and understanding by answering End of topic questions

Resources

Student Book pages 275–276

Approach

Introduce the activity

Ask students to work with a partner to make a list of key words from this topic. They could then work together to produce a spider diagram showing how the different concepts are linked. They could compare their list with the list of key words given on page 275 of the Student Book. Discuss the checklist given on page 275 and use questioning to see how much of the content students are comfortable with.

Develop the activity

Ask students to work individually to work through the End of topic questions on page 276 of the Student Book without reference to the text. As they work, walk around the classroom observing their answers and questioning them as necessary to find out which questions are causing difficulties.

Bring the activity together

After a set period, ask the students to stop working and discuss any areas of difficulty you observed as you walked round the class. Students should complete any unanswered questions for homework, but you should stress that they should attempt the questions without reference to the text to see how much content they have retained.

Answers

End of topic questions mark scheme

Question	Correct answer	Marks
1 a)	solids keep their shape because their particles are closely packed together and are held in a rigid shape by the bonds between the particles	1
	whereas the particles in liquids and gases are able to move	1
1 b)	gases fill their container because their particles move about freely,	1
	whereas the particles in a solid vibrate within their structure and those in a liquid only slide over each other	1
2	evaporation	1
3	mass	1
	temperature rise	1
	specific heat capacity	1
4	during evaporation only a few particles have enough energy to escape from the liquid state to the gas state	1
	in boiling, all the particles	1
	have sufficient energy to escape	1
5	in a solid the particles are in fixed positions	1
	the particles gain energy and the bonds between them are broken	1
	they gain enough energy so that they can move freely in a liquid	1
6	$\Delta Q = m \times c \times \Delta T$	1
	$c = \frac{\Delta Q}{m \times \Delta T}$	2
	$c = \frac{80\,320}{2 \times 80}$	3
	$c = 502$	4
	J/kg °C	1
7	the windows are cold	1
	the water vapour molecules collide with the window	1
	the window absorbs their energy and they stay as a liquid	1
8	liquid B will evaporate more easily	1
	because less energy has to be absorbed	1
	to break the bonds in liquid B	1

Question	Correct answer	Marks
9	temperature of the water stays at 100 °C until all the water has evaporated, because once boiling point reached, any additional energy transferred to the water does not increase its temperature but instead is used to change the water from a liquid to a gas (steam)	1 1 1
10	$\Delta Q = m \times c \times \Delta T$ $\Delta Q = 250 \times 4200 \times 30$ ΔQ = 31 500 000 J or 31.5 MJ	1 1
11	$\Delta Q = m \times c \times \Delta T$ $\Delta T = \Delta Q \div (m \times c)$ $\Delta T = 892\,500 \div (2.5 \times 4200)$ ΔT = 85 °C Boiling temperature = 100 °C So cold water temperature = 100 − 85 = 15° C	1 1 1 1
12	lower air pressure means less force pushing down on the surface of a liquid this means a liquid particle needs less energy in order to escape from the surface, so the liquid will evaporate more quickly	1 1 1
	Total:	37

P5d Ideal gas molecules

Introduction

In this topic students will learn about the ideal gas laws. They will learn how the Kelvin scale of temperature is derived. The content of this topic is theoretical and does not lend itself to practical work. Some students may find this difficult.

Links to other topics

Section	Essential background knowledge	Useful links
5 Solids, liquids and gases	P5b Density and pressure	P5c Change of state

Topic overview

P5d.1	**Kinetic theory** In this activity students will learn about the kinetic theory of gases, and be introduced to the three experimental gas laws. They will learn about Brownian motion and how it is supporting evidence for the theory.
P5d.2	**The Kelvin temperature scale** In this activity students will learn about the Kelvin scale of temperature and how the temperature of a gas is proportional to the average kinetic energy of its molecules.
P5d.3	**Pressure, temperature and volume** In this activity students will learn the relationships between pressure, temperature and volume for an ideal gas.
P5d.4	**Consolidation and summary** This activity provides an opportunity for a quick recap on the ideas encountered in the topic as well as time for the students to answer the End of topic questions in the Student Book.

Activity P5d.1 Kinetic theory

Specification reference: P5d.5.15

Learning objectives

- Explain how molecules in a gas have a random motion and that they exert a force and hence a pressure on the walls of the container

Learning outcomes

- Be able to use the kinetic theory to describe solids, liquids and gases
- Be able to apply the kinetic theory in a range of situations, using it to explain the observed behaviour of gases in particular

Resources

Student Book pages 277–279

Approach

Introduce the topic

Devise a quick quiz to check that students have the prior knowledge set out on page 277 of the Student Book. Then ask students to explain how particles are arranged in solids, liquids and gases.

Develop the activity

Discuss the content of pages 277–278 of the Student Book. Take time to ensure that students understand how the observed features of a gas are related to ideas from kinetic theory. Ask students to work in pairs to produce their own summary of this part of the activity.

Discuss the ideas on page 278 of the Student Book headed 'Temperature and pressure of a gas'. Ask students to discuss the questions on page 279 (top) in pairs.

Now move on to consider the experimental results listed on page 279 of the Student Book. This content will be important later in the topic so do not rush. Ask students to discuss the questions on page 279 (bottom) of the Student Book to check their understanding.

Bring the activity together

Ask students to work with a partner to produce a spidergram to summarise the main points from this activity. You may wish pairs to combine to compare notes and add to their spidergrams if necessary.

Answers

Page 279 (top)

1. The particles are colliding with the walls of the container.

2. Faster molecules hit the walls harder, so create a bigger force.

Page 279 (bottom)

1. The molecules are always moving about and spread out throughout the container.

2. The pressure of the gas increases as temperature increases.

3. A higher temperature means the molecules move more quickly, so the force on the walls will be higher.

Activity P5d.2 The Kelvin temperature scale

Specification reference: P5d.5.16, P5d.5.17, P5d.5.18, P5d.5.19

Learning objectives

- Understand why there is an absolute zero of temperature which is –273 ^{0}C
- Describe the Kelvin scale of temperature and be able to convert between the Kelvin and Celsius scales
- Understand why an increase in temperature results in an increase in the average speed of gas molecules
- Know that the Kelvin temperature of the gas is proportional to the average kinetic energy of its molecules

Learning outcomes

- Be able to use the Kelvin scale of temperature

Resources

Student Book pages 280–281

Approach

Introduce the topic

Ask students whether they think it is possible for particles in a gas to be motionless. Give them a few minutes to discuss this in small groups and then bring the class back together to discuss responses. Ask students to justify their ideas.

Develop the activity

Discuss the content on pages 280–281 of the Student Book, which answers the question posed at the beginning of the activity. Make sure that students understand that an increase in temperature results in an increase in the average speed of gas molecules and vice versa, and this is why there is an absolute zero of temperature.

Ask students to draw up a conversion table between degrees Celsius and Kelvin. Then check that they can convert between the two scales using the questions on page 280 of the Student Book.

Discuss the ideas on energy and temperature on page 281 of the Student Book. Ask students what will happen to the temperature if the average kinetic energy of the molecules is divided by three (it will divide by three).

Bring the activity together

Ask students to work with a partner to produce a poster to summarise the main points from this activity.

Answers

Page 280

1. a) 323 K.

   b) 2273 K.

2. a) –173 °C.

   b) 1227 °C.

3. They stop moving/vibrating.

Activity P5d.3 Pressure, temperature and volume

Specification reference: P5d.5.20, P5d.5.21, P5d.5.22

Learning objectives

- Explain the qualitative relationships between pressure and volume at constant temperature and between pressure and Kelvin temperature at constant volume
- Use the relationship between the pressure and Kelvin temperature of a fixed mass of gas at constant volume: $\frac{p_1}{T_1} = \frac{p_2}{T_2}$
- Use the relationship between the pressure and volume of a fixed mass of gas at constant temperature: $p_1V_1 = p_2V_2$

Learning outcomes

- Be able to use equations to link pressure, volume and Kelvin temperature of a fixed amount of gas

Resources

Student Book pages 282–285

Resources for class demonstration: a bicycle pump

Approach

Introduce the activity

Ask students to explain the link between temperature in Kelvin and energy of the particles of a gas. Now ask them to explain why there is pressure exerted on a container by gas molecules. Ask them to predict what they think will happen to the pressure of a gas in a sealed container if the temperature of the gas increases.

Develop the activity

Follow the introduction by discussing the content in the table on page 281 about the pressure law and Boyles law. These ideas need to be appreciated beforehand so everyday examples to get the students thinking are useful. For example, 'Why are car tyre pressure measurements taken when they are cool? (i.e. pressure law). 'What happens to an inflated football or balloon if it is squashed?' (Boyles law).

Also look at page 282 of the Student Book, which looks at the link between pressure and temperature. Students should attempt the worked example on pages 282– 283 without the given solution, and then, after a short period of time, bring the class back together to compare the given solution with students' attempts and compare the two.

Discuss the reasons for any differences. At this point, students can work individually or in pairs on the questions on page 283 which are designed to check understanding of the content covered so far.

If possible, show students a bicycle pump and operate its piston. Give students the opportunity to feel what happens to the piston if the air outlet is blocked.

Ask students to attempt to explain the observation in terms of the molecules of the gas. After a short time, use the explanation given in the Student Book and discuss it. Students should now attempt the worked example on page 285, before working on the questions which follow to check their understanding.

Bring the activity together

Students should work in pairs to construct a summary of the main points of the activity. These summaries should include the equations showing relationships between pressure and temperature and pressure and volume. Are students able to think how these equations might be combined?

Discuss any points that students had difficulty with in this activity.

Answers

Page 283

1. The temperature is proportional to the average kinetic energy, as long as the temperature is measured in kelvin.

2. Any relevant units can be used for pressure, as long as the same units are used throughout.

3. a) 289 K.

   b) 318 kPa.

Page 285

1. The pressure is inversely proportional to volume.

2. The mass of the gas must remain constant (that is, no particles move in or out of the system). The temperature must be measured using the Kelvin scale. The gas must be ideal (not liquefy or solidify).

3. 519.8 cm^3.

Activity P5d.4 Consolidation and summary

Learning objectives

- To review the learning points of the topic
- To test understanding through answering questions

Learning outcomes

- Be familiar with the knowledge and understanding summarised in the End of topic checklist
- Be able to apply this knowledge and understanding by answering End of topic questions

Resources

Student Book pages 286–288

Approach

Introduce the activity

Ask students to work with a partner to make a list of key words from this topic. They could then work together to produce a spider diagram showing how the different concepts are linked. They could compare their list with the list of key words given on page 286 of the Student Book. Discuss the checklist given on page 286 and use questioning to see how much of the content students are comfortable with.

Develop the activity

Ask students to work individually to work through the End of topic questions on pages 287–288 of the Student Book without reference to the text. As they work, walk around the classroom observing their answers and questioning them as necessary to find out which questions are causing difficulties.

Bring the activity together

After a set period, ask the students to stop working and discuss any areas of difficulty you observed as you walked round the class. Students should complete any unanswered questions for homework, but you should stress that they should attempt the questions without reference to the text to see how much content they have retained.

Answers

End of topic questions mark scheme

Question	Correct answer	Marks
1	as temperature increases, speed increases – it is NOT a proportional link. (If the Kelvin scale of temperature is used, then the average **kinetic energy** of the molecules is proportional to the temperature.)	1 1
2	the molecules, in random motion, hit the container walls this causes a force to act on the walls (look back to ideas included in Newton's second law) and pressure = force/area	1 1
3	it is the lowest possible temperature, the temperature where the molecules would have zero kinetic energy (and zero speed)	1 1
4 a) i)	293 K	1
4 a) ii)	423 K	1
4 a) iii)	1273 K	1
4 b) i)	27 °C	1
4 b) ii)	377 °C	1
4 b) iii)	727 °C	1
5	$V_2 = V_1T_2/T_1 = (1500 \times 260)/293$ = 1330 cm^3	1 1 1
6	$p_2 = p_1V_1/V_2 = 10^5 \times 20/5$ $= 4 \times 10^5$ Pa	1 1 1
7	$p_2 = p_1T_2/T_1 = 5 \times 10^5 \times 325 \div 290$ $= 5.6 \times 10^5$ Pa	1 1 1
8	$V_2 = p_1V_1/p_2 = 1.5 \times 10^5 \times 5 \times 10^{-3} \div 1 \times 10^5$ = 0.075 (or 7.5×10^{-3}) m^3	1 1 1
9	$T_2 = p_2T_1/p_1 = 5 \times 10^6 \times 290 \div 2.5 \times 10^6$ = 580 K	1 1 1

Question	Correct answer	Marks
10 a) i)	1.08 m^3	1 1
10 a) ii)	0.91 m^3	1 1
10 a) iii)	910 minutes	1 1
10 b)	$T_2 = p_2T_1/p_1 = 1 \times 10^6 \times 280 \div 9.5 \times 10^5$ = 295 K	1 1 1
	Total:	36

Section 5: Answers to exam-style questions

Question	Correct answer	Marks
2 a)	density = mass ÷ volume	1
2 b) i)	need small scale divisions so use ruler marked in mm	1 1
2 b) ii)	difficult to mould clay to uniform shape	1 1
2 c)	put water to known level in measuring cylinder submerge clay in water measure new water level difference gives volume of clay	1 1 1 1
2 d) i)	set balance to zero and place on it	1
2 d) ii)	will be inaccurate	1
2 d) iii)	compare with known mass/balance that is known to be correct	1 1
4 a) i)	D	1
4 a) ii)	A	1
4 a) iii)	a maximum of two from: most heat needed to raise temperature least temperature rise per energy unit lowest gradient	1 1 1
4 b) i)	water would freeze	1
4 b) ii)	idea of reducing chance of burns / risk to people	1
4 c) i)	1000 × 10 × 60 = 600 000 (J) OR 600 kJ	1
4 c) ii)	680 000 = 10 × 1700 × 40	1
4 c) iii)	the idea that energy is being transferred away while the radiator is warming up: radiator is radiating energy as it warms up/extra energy must be supplied thermal energy radiated to room the idea that the radiator itself must change its temperature: radiator case / metal needs heating (in addition to liquid) temperature rise is 32 (°C)	1 1 1 1
	Total	26

Section 6. Magnetism and electromagnetism

Contents

Overview of the section

The first, fairly short, topic in this section revises some content on magnetism that students will probably have met earlier in their studies. However, this topic goes on to discuss the difference between magnetically hard and magnetically soft materials, which may be new to students, and how magnetism can be induced in some materials.

The second topic is about electromagnetism. Students learn that an electric current in a conductor produces and magnetic field around it and they are introduced to the construction of electromagnets. The magnetic field patterns for various current-carrying conductors are discussed. The topic then moves on to explain that a force is exerted on a current-carrying conductor in a magnetic field, and students see how this effect is applied in simple d.c. motors, etc.

The final topic considers electromagnetic induction. Students learn how electricity is generated by the rotation of a magnet in a coil of wire or by a coil of wire rotating in a magnetic field. The structure of transformers is discussed and their use in the large-scale generation and transmission of electricity is explained. The equations used for ideal transformers are described and used.

Starting points

The Student Book provides a 'section opener', a double-page spread which sets the scene for students (see pages 292–293). It is structured in this way:

- an overview providing details of the areas of study
- six questions, the first three relating to aspects of physics that should be reasonably familiar to most students, and the final three relating directly to some of the topics to be studied
- a list of section contents.

The questions provide a structure for introducing the section.

1. How can electromagnets be made stronger?
2. Describe some similarities and differences between magnets and electromagnets.
3. Describe the form of the electromagnetic field around the Earth.
4. What is the difference between a magnetically 'hard' and a magnetically 'soft' material?
5. How could you investigate the magnetic field pattern for (a) a permanent bar magnet and (b) the field between two bar magnets?
6. What is a solenoid?

The section opener has two main purposes:

- to acknowledge the student's prior learning and to value it
- to provide a benchmark against which future learning can be compared.

The six questions or 'starting points' in the Student Book can be used in a number of different ways to introduce the section:

- You could ask students to consider the questions as an introductory homework task.
- You could put students into groups to share their own ideas and understanding and then report back to the whole class.
- Students could be given access to the Internet, preferably with a tight timescale, to find out the information required.

You could then use a spider chart or other form of wall chart to summarise everybody's ideas. Recording these initial ideas in this way allows you to retain them for reference as the individual topics are developed. In this way, your students' progress in learning can be readily acknowledged.

P6b Magnetism

Introduction

In this topic students will explore magnetism. They will explore magnetically hard and soft materials. They will be introduced to the concept of a magnetic field and how it can be represented by field lines, and to inducing magnetism in some materials when they are placed in a magnetic field.

Links to other topics

Section	Essential background knowledge	Useful links
6 Magnetism and electromagnetism		P6c Electromagnetism P6d Electromagnetic induction

Topic overview

P6b.1	**Introduction to magnetism** In this activity, students will learn that magnets repel and attract other magnets and attract magnetic substances. They will be introduced to magnetically hard and soft materials and the concept of a magnetic field and how it can be represented by magnetic field lines.
P6b.2	**Magnetic field patterns and induction** In this activity students will first be shown how magnetism can be induced in some materials when they are placed in a magnetic field and then they will explore magnetic field patterns.
P6b.3	**Consolidation and summary** This activity provides an opportunity for a quick recap on the ideas encountered in the topic as well as time for the students to answer the End of topic questions in the Student Book.

Activity P6b.1 Introduction to magnetism

Specification reference: P6b.6.2, P6b.6.3, P6b.6.4

Learning objectives

- Know that magnets repel and attract other magnets and attract magnetic substances
- Describe the properties of magnetically hard and soft materials
- Understand the term magnetic field line

Learning outcomes

- Understand that magnets repel and attract other magnets
- Understand that magnets attract magnetic substances
- Be able to describe the properties of magnetically hard and soft materials

Resources

Student Book pages 295–297

Resources for class demonstration (see Technician's notes, following)

Approach

Introduce the activity

Ask students what they know about magnets. As a class, produce a list of things that students know about magnetism.

Develop the activity

Discuss the content on page 296 of the Student Book about magnets repelling and attracting. If possible, demonstrate the alignment of a bar magnet as shown in Figure 6.2 on page 296. Students can discuss the questions on page 296 in pairs to check understanding.

Now move on to discuss magnetically hard and soft materials. Ask students to work with a partner to make a summary of this content and answer the questions on page 297.

Students can now tackle the questions on page 300, either individually or in pairs.

Bring the activity together

Ask students to work with a partner to make a summary of the key points of this activity. After a short while, bring the class together to compare their summaries with the stated learning outcomes of this activity. Discuss why the two differ if there are differences.

Technician's notes

Be sure to check the latest safety notes on these resources before proceeding.

The following resources are needed for the class demonstration P6b.1:

bar magnet, ruler, string	bar magnet, compass
bar magnet, paper, iron filings	eye protection

If you decide to ask students to plot the magnetic field lines around a bar magnet, you will need a bar magnet, compass, sheet of paper and pencil for each group.

Answers

Page 296

1. They attract each other.
2. They repel each other.
3. It will line up approximately north–south.

Page 297

1. Magnetically hard materials retain their magnetism; magnetically soft materials lose their magnetism if the outside magnetic influence is removed.
2. A permanent magnet that is made of magnetically hard materials.
3. In electromagnets and relays.

Activity P6b.2 Magnetic field patterns and induction

Specification reference: P6b.6.4, P6b.6.5, P6b.6.6, P6b.6.7

Learning objectives

- Understand the term 'magnetic field line'
- Know that magnetism is induced in some materials when they are placed in a magnetic field
- Describe how to use two permanent magnets to produce a uniform magnetic field pattern

Learning outcomes

- Know how to plot magnetic fields
- Understand that magnetism is induced in some materials when they are placed in a magnetic field
- Describe experiments to investigate the magnetic field pattern for a permanent bar magnet and that between two bar magnets
- Be able to use the idea of magnetic field lines to explain magnetic effects

Resources

Student Book pages 297–303

Worksheet P6b.2a Finding magnetic field patterns

Resources for class demonstration and practical (see Technician's notes, following)

Approach

Introduce the activity

Demonstrate magnetic induction as shown in Figure 6.8 of the Student Book. Ask students how this demonstration helps to distinguish between a soft magnetic material and a permanent magnet. Students can answer the questions at the bottom of page 300 of the Student Book individually or in pairs.

Develop the activity

If you have sufficient equipment, allow students to work in small groups to investigate the magnetic field patterns shown in Figure 6.9 of the Student Book. (The instructions for doing this can be found on Worksheet P6b.2a.) If you do not have sufficient equipment, then do this as a class demonstration, making sure that students are involved at all stages: telling you what to do next, plotting the lines, etc.

Now demonstrate how to find the field lines around a bar magnet, as shown in Figure 6.4 on page 297 of the Student Book. You could also demonstrate this by using a compass to plot the field lines. Ask students to give you instructions on to how to do this. The Student Book outlines the procedure. You could also ask students to carry out this activity themselves.

Students should then tackle the questions at the top of page 300 of the Student Book.

Now consider how to produce a uniform magnetic field. Demonstrate the uniform magnetic field between the poles of a horseshoe magnet. Students should then answer the questions on page 303 of the Student Book.

Bring the activity together

Students should cover the Science in context box on page 298. This can be followed by all the questions on page 302. After this, they should make a poster to summarise the learning in this activity. This should include diagrams to show the magnetic field patterns that they have explored.

Technician's notes

Be sure to check the latest safety notes on these resources before proceeding.

The following resources are needed for the class demonstration P6b.2:

bar magnet	iron filings
piece of soft magnetic material	paper
horseshoe magnet	eye protection

The following resources are needed for the class practical P6b.2a, per group:

bar magnet	plotting compass
iron filings	eye protection
thin sheet of plastic	

Answers

Page 300 (top)

1. A region of space where their magnetism affects other objects.
2. The path that a free north pole would take from a north pole to a south pole.
3. Using iron filings on a thin sheet of plastic or a plotting compass.

Page 300 (bottom)

1. It has magnetism induced in it and is attracted to the magnet.
2. A south pole.
3. Attraction only tells you the material is a magnetic material. Only two magnets can repel, so that is the test.
4. Two north poles together or two south poles together.

Page 302

1. The magnetic field lines radiate out from the north pole and go round to the south pole.
2. They attract each other and the field lines go from a north pole to a south pole.
3. The field lines go from a north pole to a south pole, not between like poles.

Page 303

1. Magnetic field lines show the direction of a force (on a free north pole); if field lines crossed it would indicate a force in more than one direction, which makes no sense.
2. Even, the same strength at all places, etc.
3. Between the poles of a U-shaped magnet, or when opposite poles of two magnets are placed close to each other.

P6b.2a Finding magnetic field patterns

In this practical you will find magnetic field patterns.

Apparatus

bar magnet

thin sheet of plastic iron filings

plotting compass

eye protection

SAFETY INFORMATION

Iron filings are an irritant.

Avoid contact with eyes.

Wear eye protection.

Wash hands after use.

Take care not to get iron filings on the magnet.

Method

1. Place the bar magnet under a thin sheet of plastic.
2. Sprinkle iron filings on to the top of the plastic.
3. Follow the path of the field lines by placing a small compass (known as a plotting compass) on the plastic in place of the iron filings.

Analyse and interpret data

1. Draw the field lines around the bar magnet.
2. What happens if you move the compass in the direction that its north pole is pointing?

Evaluate data and methods

1. Why do you need a thin sheet of plastic rather than paper?
2. What would happen if you used too many or too few iron filings?

Activity P6b.3 Consolidation and summary

Learning objectives

- To review the learning points of the topic
- To test understanding through answering questions

Learning outcomes

- Be familiar with the knowledge and understanding summarised in the End of topic checklist
- Be able to apply this knowledge and understanding by answering End of topic questions

Resources

Student Book pages 304–305

Approach

Introduce the activity

Ask students to work with a partner to make a list of key words from this topic. They could then work together to produce a spider diagram showing how the different concepts are linked. They could compare their list with the list of key words given on page 304 of the Student Book. Discuss the checklist given on page 304 and use questioning to see how much of the content students are comfortable with.

Develop the activity

Ask students to work individually to work through the End of topic questions on page 305 of the Student Book without reference to the text. As they work, walk around the classroom observing their answers and questioning them as necessary to find out which questions are causing difficulties.

Bring the activity together

After a set period, ask the students to stop working and discuss any areas of difficulty you observed as you walked round the class. Students should complete any unanswered questions for homework, but you should stress that they should attempt the questions without reference to the text to see how much content they have retained.

Answers

End of topic questions mark scheme

Question	Correct answer	Marks
1	a magnetically hard material stays magnetic once it has been magnetised	1
	example: steel	1
	a magnetically soft material does not stay magnetic when the source of magnetisation is removed	1
	example: pure iron	1
2	the piece of metal could just be made of a soft magnetic material	1
	this will be attracted if it is brought near to a magnet	1
	Ranjit could turn the piece of metal around and see if it is still attracted (in which case it is not a magnet) or repelled (in which case it is a magnet)	1
3	see page 301 for the magnetic field pattern for a single bar magnet	1
	if the magnet were made stronger, the field pattern would not change	1
	but it is common to indicate the increased field strength by drawing more lines between the original ones	
4	place the bar magnets under a sheet of paper	1
	sprinkle iron filings on top of the paper	1
	the iron filings will show the field pattern of each magnet	1
	and the pattern between the magnets	1
5	take two magnets of equal strength	1
	place them with opposite poles facing each other	1
	the magnetic field between the poles will be uniform	1
	the same effect can be seen between the poles of a horseshoe magnet	1
	Total:	17

P6c Electromagnetism

Introduction

In this topic students will learn that an electric current in a conductor produces a magnetic field around it. They will learn about electromagnets before moving on to consider how a force on a current-carrying wire can be used in devices such as motors.

Links to other topics

Section	Essential background knowledge	Useful links
2 Electricity		P2c Energy and voltage in circuits P2d Electric charge
3 Waves		P3c The electromagnetic spectrum
6 Magnetism and electromagnetism	P6b Magnetism	P6d Electromagnetic induction

Topic overview

P6c.1	**Electromagnets** In this activity students will learn that an electric current in a conductor produces a magnetic current round it. They will learn about the construction of electromagnets and the magnetic field patterns for a straight wire, a flat circular coil and a solenoid.
P6c.2	**Force on a charged particle and current-carrying wire** In this activity students will learn that a charged particle and a current-carrying wire experience a force when placed in a magnetic field. They will learn how to predict the direction of this force and how it is used in devices such as simple motors.
P6c.3	**Consolidation and summary** This activity provides an opportunity for a quick recap on the ideas encountered in the topic as well as time for the students to answer the End of topic questions in the Student Book.

Activity P6c.1 Electromagnets

Specification reference: P6c.6.8, P6c.6.9P, P6c.6.10P

Learning objectives

- Know that an electric current in a conductor produces a magnetic field round it
- Describe the construction of electromagnets
- Draw magnetic field patterns for a straight wire, a flat circular coil and a solenoid when each is carrying a current

Learning outcomes

- Be able to describe how to make an electromagnet
- Be able to sketch the magnetic field patterns for wires and coils of various shapes

Common misconceptions

The concept of a magnetic field is problematic to some students. Those who have difficulty working in the abstract will find visualisation difficult. Therefore, spend time with these students plotting the magnetic field shapes around a range of current-carrying conductors. This will help students to 'see' the magnetic field shapes, which will make the associated theory more approachable.

Resources

Student Book pages 306–310

Resources for class demonstration (see Technician's notes, following)

Approach

Introduce the activity

Show students a length of wire and ask whether it has a magnetic field around it. [No]

Now connect the wire to a battery, with a switch in the circuit, as shown in Figure 6.14 of the Student Book. Have the switch open at this stage. Ask whether the wire has a magnetic field around it now. Demonstrate, using the compass, that while the switch is open and no current is flowing, there is no magnetic field around the wire.

Ask students what they think will happen to the needle on the compass if you close the switch. Close the switch and check whether their prediction was correct.

Develop the activity

Discuss the content of pages 306–307 of the Student Book. If possible, you could demonstrate how the strength of the magnetic field around a coil of wire increases if there is an iron core in the coil (a nail is suitable for use as a core). Make sure that students see that there is only a magnetic field around the wire when there is a current flowing through it.

Allow students to work in pairs on the Developing investigative skills feature on page 308 of the Student Book. After a period of time, bring the class back together to discuss their responses to the questions, and compare these with the model responses. Discuss any differences that occur, before allowing students to work with a partner to answer the questions on page 307 which are designed to check understanding of the content so far.

The next part of the activity concentrates on field patterns around various electromagnets. You could set up a flat circular coil as shown in Figure 6.18 and involve students in plotting the magnetic field around it using compasses, as they did in Topic P6b.

Students work in pairs to answer the questions on page 310 of the Student Book to check understanding.

Bring the activity together

Prepare cards which show various magnetic field patterns an ask students to identify whether the pattern is for a straight wire, a flat circular coil or a solenoid, all of which are carrying current.

Ask students to explain the necessary conditions for a piece of wire to produce a magnetic field. Ask them to explain how to construct an electromagnet.

Technician's notes

Be sure to check the latest safety notes on these resources before proceeding.

The following resources are needed for the class demonstration P6c.1:

length of wire	nail
battery	flat circular coil, battery, compass, piece of card
switch	cards showing different magnetic field patterns
compass	

Answers

Page 307

1. Make the wire into a coil. Connect one terminal of the battery to the switch and one end of the coil of wire. Connect the other end of the coil to the switch. Connect the switch to the other terminal of the battery.

2. Increase the current, increase the number of turns on the coil.

3. Diagram should be similar to Fig. 6.16.

Developing investigative skills, page 308

1. To vary the current, the student can place a variable resistor in series with the coil and the power supply. Different settings on the variable resistor will allow different currents in the circuit. The power supply the student uses may have a variable setting to do this. The student will need to include an ammeter in the circuit to measure the current.

2. The student should keep the number of turns on the coil constant (including keeping the spacing of the turns constant). Clearly the student should use the same nail throughout as different nails may have different diameters, etc. They should also keep the temperature of the coil constant (see question 7).

3. Line graph or bar chart. The current is a continuous variable, indicating that a line graph would be appropriate. However, the discrete nature of the paper clips measurement suggests that a best fit line may not be entirely suitable in this context.

4. As the current increases, the electromagnet can generally hold more paper clips. The 'strength' of the electromagnet is not a variable on the graph, so really should not be included in a statement of 'what the graph shows'.

5. a) Hanging the paper clips in a bunch has the advantage of attaching quite a large number of paper clips, so in effect smaller variations in strength can be noted. However, it is not a particularly repeatable method since there are many angles that the clips can attach in and the student would have to judge how far up the nail to allow the paper clips to go.

b) Hanging the clips in a line attached to each other is really testing the strength of the electromagnet at its tip since the other paper clips are really just adding extra weight. This method is not a particularly fine discriminator, since it will only pick up changes in terms of 'whole paper clips'. Since the paper clips are attached, they could all be attached to the first one which should be connected to the tip of the electromagnet through magnetic attraction alone.

c) The method really tests how well the magnet induces magnetism in the first paper clip and, consequently, how each paper clip in turn induces magnetism in the next. This method will work well if the paper clips are small so that finer adjustments can be made.

6. Still using paper clips, the measurements will be more precise if smaller paper clips are used, since this will allow smaller changes in the magnetic field to be noted. If a different measuring technique is used, such as measuring the repulsion between the electromagnet and a bar magnet resting on an electronic balance (where the repulsion gives an increased reading on the balance scale) then improved precision can be achieved.

7. As the current increases the heating effect in the coil will increase. With the plastic-coated wire typically used in school science labs this can result in the plastic melting, making the coil difficult to use. Also, the low voltage power supply used may have a cut out to prevent large currents being drawn from it.

Page 310

1. Sketch should match Fig. 6.20. Similar shape to bar magnet pattern – 'fanning out' at the poles; 'loops' along the sides, etc.

2. The field at the centre of the coil is most nearly uniform.

3. a) Sketch should match right-hand part of Fig. 6.19.

   b) Sketch should match left-hand part of Fig. 6.19.

Activity P6c.2 Force on a charged particle and current-carrying wire

Specification reference: P6c.6.11P, P6c.6.12, P6c.6.13, P6c.6.14

Learning objectives

- Know that there is a force on a charged particle when it is moved in a magnetic field as long as its motion is not parallel to the field
- Understand why a force is exerted on a current-carrying wire in a magnetic field, and how this effect is applied in simple d.c. electric motors and loudspeakers
- Use the left-hand rule to predict the direction of the resulting force when a wire carries a current perpendicular to a magnetic field
- Describe how the force on a current-carrying conductor in a magnetic field changes with the magnitude and direction of the field and current

Learning outcomes

- Understand that there is a force on a charged particle when it moves in a magnetic field as long as its motion is not parallel to the field
- Be able to explain how simple d.c. motors and loudspeakers work
- Know how to predict the direction of the force on a current-carrying wire perpendicular to a magnetic field
- Know how to increase the force on a current-carrying conductor

Resources

Student Book pages 311–316

Worksheet P6c.2a Building an electric motor

Files on CD-ROM: P6c.2a_worksheet

Resources for class demonstration and practical (see Technician's notes, following)

Approach

Introduce the activity

Remind students what they learnt in the last activity about magnetic fields around current-carrying conductors. Then ask students what a current consists of (moving electrons).

Develop the activity

Explain that, if a charged particle moves in a magnetic field (as it does in a current), it experiences a force. Therefore, if a current-carrying wire is placed in a magnetic field, it experiences a force, which we call the motor effect.

Demonstrate the motor effect as follows. Connect the foil strip horizontally between a pair of clamp stands, under slight tension. Attach the wires to each end of the strip so that they are at right angles to it. Hold the horseshoe magnet so that the strip is between its poles and the two wires are perpendicular to the field direction. Attach to the power pack, set to 6 V.

Switch the power pack on and demonstrate the motor effect. Show students how to predict the direction of the force, using Fleming's left-hand rule (see page 313 of the Student Book).

Demonstrate changes to the apparatus – increasing/decreasing the current, using longer/shorter pieces of foil (will not change force), using a strong/weak magnet, reversing polarity of power pack connections.

If the dimensions of the horseshoe magnet allow, show the effect on the foil mounted parallel to the field line direction. If the apparatus does not allow this, explain the effect.

Only switch on for a few seconds as the wires and aluminium foil will get very hot.

Stress to students that the 'wire' being used is a piece of aluminium foil – a non-magnetic material. This reinforces two key concepts: the induction of a magnetic field around a current-carrying conductor and, hence, the ability to be able to switch on and off a field through the use of an electric current.

Now move on to how this effect is used. Ask students to work with a partner to make a list of as many devices as possible which use electric motors. After a short period of time, bring the class back together to discuss ideas.

At this point, if resources allow, students can carry out Worksheet P6c.2a, which involves students in making a motor. A range of motor kits are commercially available. Ensure that students spend sufficient time producing a quality motor in the first instance. Changes to the factors will not produce the desired effect if a motor is barely turning.

Discuss how a loudspeaker works using page 312 of the Student Book for guidance. Students should work in pairs to answer the questions on pages 313 and 316 to check understanding.

Bring the activity together

Ask students to explain to their working partner how an electric motor produces a rotation. Then ask students to work in pairs to produce a summary of the main points of this activity.

After a period of time, bring the class back together to discuss the summaries and ensure that they have covered the stated learning outcomes for this activity, especially the left-hand rule and factors that change the force on a current-carrying conductor.

Technician's notes

Be sure to check the latest safety notes on these resources before proceeding.

The following resources are needed for the class demonstration P6c.2:

strong horseshoe magnet
weak horseshoe magnet
long strip of aluminium foil (approximately 100 cm × 5 cm)
power pack (Weir pack, for example)
switch
demonstration ammeter (or digital multimeter)
connecting wires
crocodile clips
clamp stands

The following resources are needed for the class practical P6c.2a, per group:

motor kit
power pack or battery/cells
Magnadur magnets
multimeter

Answers

Page 313

1. The current in the coil creates a magnetic field around it. This interacts with the magnetic field from the permanent magnet, producing forces in opposite directions on either side of the coil. Putting the coil on an axle allows these forces to spin the coil.

2. Connect a length of wire across a power supply and coiled around one pole of a permanent magnet. Switch the current on and the wire jumps off the magnet.

3. The changing current in the coil produces a changing magnetic field which interacts with the field from the permanent magnet. This creates a 'backwards and forwards' motion of the coil and paper cone. This makes the air vibrate – a sound wave.

Page 316

1. The thumb, and first and second fingers are held at right angles. The first finger points in the direction of the field, the second finger in the direction of the current and the thumb gives the direction of the force (movement).

2. The commutator reverses the direction of the current in the coil each half-turn. This allows the coil to keep turning in the same direction.

3. Increase the current, increase the strength of the permanent magnets, increase the number of turns on the motor coil. (Reducing the friction on the axle would make the motor spin faster, but it doesn't increase the force, so it doesn't answer the question.)

4. Reverse the direction of the current (by turning the battery round, etc), reverse the polarity of the magnets.

P6c.2a Building an electric motor

You are going to build a simple electric motor.

Apparatus

motor kit

power pack or battery/cells

Magnadur magnets

multimeter

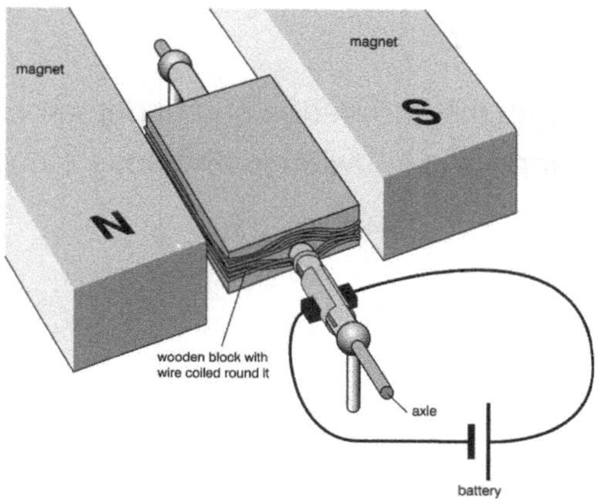

Method

1. Build your motor according to the motor kit's instruction sheet.
2. Attach a battery or power pack (3 V d.c.) to the motor. Observe the speed it rotates at.
3. Try increasing the voltage (up to maximum of 4.5 V) to increase the current supplied to the motor. What happens to the motor?
4. Make the fixed magnetic field stronger. What happens to the motor?
5. Reverse the electrical connections. What happens to the motor?
6. Remove the rotor and add more coils of wire. What happens to the motor?
7. Record your observations in the appropriate part of the Worksheet.

Recording your data

Record your observations in a table like the one below:

Factor changed	Effect on motor speed
Increase current	
Increase magnetic field strength	
Change polarity of connections	
Increase number of coils	

Analyse and interpret data

How can the force produced from an electric motor be increased?

Activity P6c.3 Consolidation and summary

Learning objectives

- To review the learning points of the topic
- To test understanding through answering questions

Learning outcomes

- Be familiar with the knowledge and understanding summarised in the End of topic checklist
- Be able to apply this knowledge and understanding by answering End of topic questions

Resources

Student Book pages 317–319

Approach

Introduce the activity

Ask students to work with a partner to make a list of key words from this topic. They could then work together to produce a spider diagram showing how the different concepts are linked. They could compare their list with the list of key words given on page 317 of the Student Book. Discuss the checklist given on page 317 and use questioning to see how much of the content students are comfortable with.

Develop the activity

Ask students to work individually to work through the End of topic questions on pages 318–319 of the Student Book without reference to the text. As they work, walk around the classroom observing their answers and questioning them as necessary to find out which questions are causing difficulties.

Bring the activity together

After a set period, ask the students to stop working and discuss any areas of difficulty you observed as you walked round the class. Students should complete any unanswered questions for homework, but you should stress that they should attempt the questions without reference to the text to see how much content they have retained.

Answers

End of topic questions mark scheme

Question	Correct answer	Marks
1	bring up a north pole from another magnet if it repels you have found the north pole of your electromagnet if it attracts then the pole is a south pole	1 1 1 1
2	increasing the current flowing through the coil increasing the number of coils adding a soft iron core inside the cardboard tube	1 1
3	when the switch is pressed, the electromagnet attracts the hammer support and the hammer hits the bell the movement of the hammer support breaks the circuit and so the electromagnet ceases to operate the hammer support then returns to its original position, forming the circuit again and the process is repeated	1 1 1 1
4	increase the current increase the number of turns on the coil	1 1
5 a)	there is no force on the particle and it keeps moving in a straight line	1
5 b)	there is a force on the particle which will push it off the straight line	1
6	greater force applied to rotor	1 1
7	several small coils of wire create same effect as one large coil but take up less space	1 1
8	yes if the magnets are the same strength	1 1
9	in a loudspeaker, the coil is attached to a paper cone the changing current in the coil produces a changing magnetic field which interacts with the field from the permanent magnet this creates a 'backwards and forwards' motion of the coil and paper cone this makes the air vibrate – a sound wave	1 1 1 1 1 1
10 a) i)	the field line is a circle, centred on the wire and passing through P the arrow on the wire should point anticlockwise	1 1
10 a) ii)	the arrow through Q should point to the left	1
10 b) i)	if the current is increased the compass will either not move at all or will move negligibly	1
10 b) ii)	if the current is reversed, the compass will point the other way	1
11	at S the field is stronger at T the field is the same at W the field is the same	1 1 1
	Total:	34

P6d Electromagnetic induction

Introduction

In this topic students will learn about electromagnetic induction. They will learn how transformers work and how step-up and step-down transformers are used in large-scale generation and transmission of electrical energy.

Links to other topics

Section	Essential background knowledge	Useful links
2 Electricity		P2b Mains electricity P2c Energy and voltage in circuits P2d Electric charge
6 Magnetism and electromagnetism	P6b Magnetism P6c Electromagnetism	

Topic overview

P6d.1	**Introduction to electromagnetic induction** In this activity students will learn how a voltage can be induced when the magnetic field around a conductor or coil changes. They will learn about the factors which affect the size of the induced voltage.
P6d.2	**Transformers** In this activity students will learn about the structure of a transformer, and the relationships between the input and output voltages and number of turns on the primary and secondary coils, and the relationship between input power and output power for 100% efficiency.
P6d.3	**Using transformers in electricity generation** In this activity students will learn about the generation of electricity and why step-up and step-down transformers are used to reduce power losses in cables.
P6d.4	**Consolidation and summary** This activity provides an opportunity for a quick recap on the ideas encountered in the topic as well as time for the students to answer the End of topic questions in the Student Book.

Activity P6d.1 Introduction to electromagnetic induction

Specification reference: P6d.6.15, P6d.6.16

Learning objectives

- Know that a voltage is induced in a conductor or a coil when it moves through a magnetic field or when a magnetic field changes through it and describe the factors which affect the size of the induced voltage
- Describe the generation of electricity by the rotation of a magnet within a coil of wire and of a coil within a magnetic field and describe the factors which affect the size of the induced voltage

Learning outcomes

- Be able to describe how electromagnetic induction is used to generate electric currents

Common misconceptions

Understanding the concept of cutting magnetic flux is difficult for many students. Use diagrams to visualise the field around a magnet and the relative position of a wire within this field. Encourage students to follow a similar approach to help them 'see' what a question is asking.

Resources

Student Book pages 320–324

Worksheet P6d.1a Electromagnetic induction

Worksheet P6d.1b Induction from a coil

Files on CD-ROM: P6d.1a_worksheet, P6d.1b_worksheet

Resources for class demonstration and practicals (see Technician's notes, following)

Approach

Introduce the activity

Ask students to sketch the shape of the magnetic field for a number of different situations – for example, around a bar magnet, a current-carrying conductor and a solenoid.

Develop the activity

Show that a voltage can be induced by dropping a magnet through a large demonstration coil. Explain that in the previous activity, a voltage was supplied to a coil, which produced a current, in a magnetic field – movement resulted. What has just been observed is the opposite effect – movement is being supplied to a coil in a magnetic field and an induced voltage results.

Show the effect of moving the magnet into and out of the coil and of changing the polarity of the magnetic field. (Refrain from demonstrating the effect of moving the magnet more quickly or carrying out other typical demonstrations, such as moving the coil relative to the magnet. These ideas will be explored during the student practical.)

If resources permit, ask students to carry out the practical on Worksheet P6d.1a, or do it as a demonstration. Allow time to discuss the results as a class.

If you wish, work on the links between induced voltage and current can be explored in more detail with the use of a demonstration coil and a data logger with voltage and current sensors attached. Drop a magnet through the coil, noting the resulting voltage and currents induced. Repeat this, adding a resistor to the circuit.

Students should note the differences in the resulting sensor readings. Avoid the use of vocabulary such as 'magnetic flux linkage' at this level – many students find this terminology confusing. Limit discussions to the concept of wires 'cutting field lines'.

Discuss the content of pages 321–322 of the Student Book and allow time for students to discuss the questions on pages 321 (two sets) and 322 with a partner.

The Developing investigative skills feature on page 323 can also be tackled in pairs – discussion of the questions can help understanding. You may wish to compare student responses to the model answers given and discuss any differences that arise.

More able students could be asked to look at the extension feature on page 324 as a homework exercise.

If resources permit, ask students to carry out the practical on Worksheet P6d.1b, or do it as a demonstration. More able students should be asked to predict the outcome of each action in the practical prior to gaining results and should justify their hypotheses. Once results have been collected, students should draw parallels between this practical and the previous practical.

Bring the activity together

Ask students to work with a partner to make a summary of the main points covered in this activity. After a period of time, bring the class together to discuss the summaries and ensure that the stated learning outcomes have been noted.

Technician's notes

Be sure to check the latest safety notes on these resources before proceeding.

The following resources are needed for the class demonstration P6d.1:

large demonstration coil
magnet
voltmeter or ammeter (large enough for demonstration)
connecting leads and crocodile clips
data logger with voltage and current sensors (optional)
resistance substitution box or variable resistor (optional)

The following resources are needed for the class practical P6d.1a, per group:

coils of wire – varying numbers of coils and cross-sectional areas
magnets of differing strengths
connecting leads and crocodile clips
voltmeters (ideally analogue centre-zero type)

The following resources are needed for the class practical P6d.1b, per group:

power packs
connecting leads and crocodile clips
voltmeters (ideally analogue centre-zero)
cardboard tubes – varying cross-sectional areas
thin wire, insulated
iron cores

Only keep coil A switched on for a few seconds at a time as the wires will get very hot.

Answers

Page 321 (top)

1. Relative motion between the magnetic field and the wire – a current will flow if there is a complete loop of wire; if there isn't, a voltage will be generated across the ends of the wire.

2. No, because the wire is not moving.

3. By changing the strength of the magnetic field and/or the speed the wire is moving at.

Page 321 (bottom)

1. A larger current is created.

2. The area of the coil, the number of turns in the coil, the strength of the magnetic field and the speed of rotation.

3. A coil of area 2 m^2.

Page 322

1. The magnet is rotated inside the coil.

2. The speed of rotation.

3. To maximise the effect, increase the rate of rotation, increase the strength of the magnetic field, increase the number of turns on the coil. The output is an alternating current because the relative motion between coil and field reverses each half-turn.

Developing investigative skills, page 323

1. The time period of the spring's oscillation depends on the stiffness of the spring, so if the student sets the spring going from larger extensions the initial speed will be greater, indicating that the method will work. However, as the oscillation loses energy and the amplitude decreases, so the speed will reduce as well. As long as the student is only trying to measure the initial maximum voltage this method will allow measurements to be made. If the student were to use different springs with different spring constants they would also be able to control the speed of the oscillation.

2. The timebase circuit moves the trace across the screen, so if it is switched off the student will see a trace that moves up and down only, reflecting the magnitude of the induced voltage.

3. The deflection of the screen will be positive as the magnet moves in one direction and negative when the magnet moves in the other direction.

4. The screen of the CRO has a grid. The controls of the CRO will indicate how many volts each grid square is 'worth'. As the trace is deflected the student will be able to judge how many grid squares it moves and so will be able to estimate the maximum voltage.

5. Graph should have a sine wave shape with an amplitude that decreases over time.

6. In this context, qualitative indicates a more descriptive measurement (bigger, smaller) rather than specific measurements (a quantitative approach).

7. The measurements of speed and voltage could be improved by developing a method that makes some kind of record that can be analysed later, rather than judging by sight as the experiment runs. A data logger and sensors could be used, or record the experiment with a video camera, allowing playback frame by frame at a later time.

Page 324

Students' own answers that should cover the following:

Equipment needed: magnet, coil of wire and sensitive meter.
Method: Vary the current by altering the number of coils and the speed the magnet is moved in the coil.

P6d.1a Electromagnetic induction

In this practical you will explore electromagnetic induction.

Apparatus

coils of wire

different strength magnets

centre-zero voltmeter

connecting leads

crocodile clips

Method

Arrange your apparatus as follows:

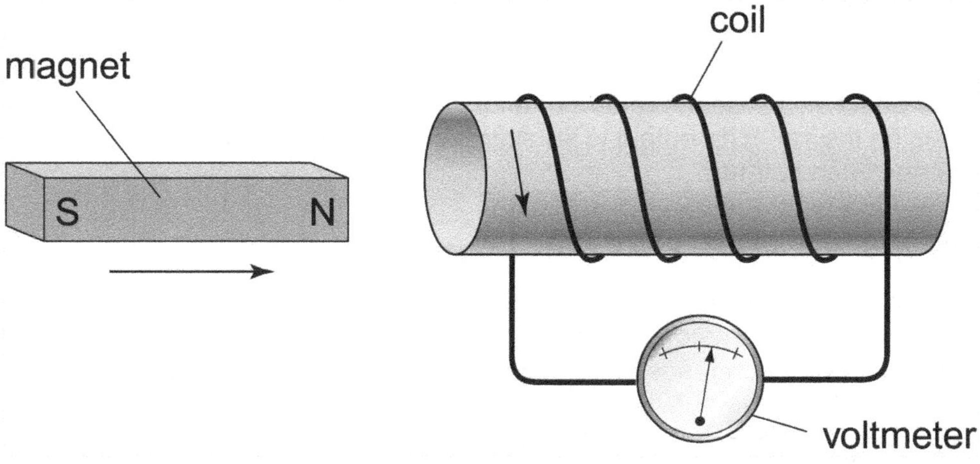

For each of the following steps, note the maximum reading on the voltmeter and record your observations in the results table below.

1. Insert the magnet into the coil.
2. Insert the magnet into the coil more quickly.
3. Repeat step 2, but use the opposite pole of the magnet.
4. Try using a stronger magnet – what is the effect now?
5. Insert the magnet into a coil with more turns of wire.
6. Insert the same magnet, at the same speed, but use a coil of greater cross-sectional area.
7. Now hold the magnet still and move the coil around close to the magnet.
8. Repeat step 7, but move the coil more quickly.

Analyse and interpret data

1. Record your observations in the table:

Action	Observation
Magnet inserted slowly into coil	
Magnet inserted quickly into coil	
Opposite pole of magnet inserted and removed from coil	
Stronger magnet inserted into coil	
Magnet inserted into coil with more turns	
Magnet inserted into coil of greater cross sectional area	
Coil waved slowly around magnet	
Coil waved quickly around magnet	

2. What do all of your observations have in common with each other?

P6d.1b Induction from a coil

In this practical, you will investigate induction from a coil.

Apparatus

insulated wire

cardboard tubes

power pack

centre-zero voltmeter

connecting leads

crocodile clips

iron cores

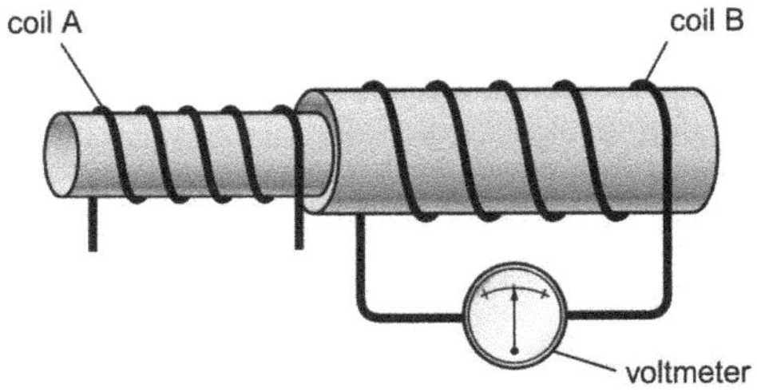

SAFETY INFORMATION

Only keep coil A switched on for a few seconds at a time as the wires will get very hot.

Method

Arrange your apparatus as shown in the diagram. Begin with 50 coils of wire on both cardboard tubes. Follow the steps below and record the maximum voltage for each result obtained in the results table on the Worksheet.

1. Connect coil A to a power pack.
2. Set the power pack to 2 V d.c.
3. Move coil A towards coil B. Note the reading on the voltmeter.
4. Pull coil A away from coil B. Note the reading on the voltmeter.
5. Keep coil A stationary and move coil B. Note the reading on the voltmeter.
6. Repeat steps 3 and 4, making the following changes:
   a) Move the coils more quickly.
   b) Add more coils onto coil A.
   c) Add more coils onto coil B.
   d) Add an iron core into each of the cardboard tubes.
   e) Increase the voltage applied to coil A.
   f) Change the supply from d.c. to a.c..

Analyse and interpret data

1. Record your results in the table:

Conditions	Maximum induced voltage (V)
Coil A moved slowly	
Coil A moved more quickly	
More coils on coil A	
More coils on coil B	
Iron core added to coils	
Increased voltage supplied to coil A	

2. When is a voltage induced in coil B?
3. How can the voltage induced in coil B be increased?
4. Is it possible to induce a voltage in coil B when both coils are stationary?
5. What do all of your observations have in common with each other?

Activity P6d.2 Transformers

Specification reference: P6d.6.17P, P6d.6.19P

Learning objectives

- Describe the structure of a transformer, and understand that a transformer changes the size of an alternating voltage by having different numbers of turns on the input and output sides
- Know and use the relationship between the input (primary) and output (secondary) voltages and the turns ratio for a transformer: $\frac{\text{input (primary) voltage}}{\text{output (secondary) voltage}} = \frac{\text{primary turns}}{\text{secondary turns}}$

  $\frac{V_p}{V_s} = \frac{n_p}{n_s}$

Learning outcomes

- Be able to describe the structure and function of a transformer
- Be able to carry out calculations relating to transformers

Common misconceptions

Many students gain the misconception that the iron core of a transformer carries an electrical current. Demonstrate clearly that insulated wires must be used on both the primary and secondary sides and hence no current could flow through the iron core.

The idea of a 'magnetic conductor' is often used as an analogy to an electrical conductor, to describe the transfer of the magnetic field between the primary and secondary sides. Ensure that students take care to separate the ideas of electrical conductance and magnetic field propagation in an explanation of how a transformer works.

Resources

Student Book pages 324–326

Practical P6d.2a Building a transformer

Files on CD-ROM: P6d.2a_worksheet

Resources for class demonstration and practical (see Technician's notes, following)

Approach

Introduce the activity

Put students into small groups. Ask each group to come up with a definition for the term 'electromagnetic induction'. Share the definitions amongst the class.

Develop the activity

Use a demonstration transformer to show the structure and concept of a transformer. Set up the demonstration transformer with 300 turns on the primary side and 600 turns on the secondary side. Connect to the d.c. side of the power pack (approx. 4 V) and switch on.

Ask students to explain why there is no voltage induced on the output side. Attach an a.c. supply of the same voltage and ask students to note the output voltage. Reverse the turns ratio and demonstrate a step-down transformer.

Simply describe the idea of being able to increase or decrease the output voltage. Ensure that students are comfortable with the vocabulary used – soft iron core, primary and secondary coils, step-up and step-down transformers, magnetic field, cutting magnetic field lines and induced voltage.

Discuss the content of pages 324–325 of the Student Book. Allow students to attempt the worked example on page 325 without the given solution and then, after a short period, compare their solutions with the model solution.

If resources allow, ask students to carry out the practical on Worksheet P6d.2a, or do it as a demonstration. Note that using C-cores can lead to significant inefficiency in a transformer.

Encourage students to take their time over collecting results and ensure that coils are wound tightly around the C-cores and that the two C-cores are held together rigidly as results are being taken. Challenge more able students to produce an output from their transformer using a d.c. supply. Summarise the results as a class.

More able students can look at the Extension feature on page 325 of the Student Book and all students should work in pairs to tackle the questions on page 326 which are designed to check understanding.

Bring the activity together

Project an image of a transformer onto the whiteboard, along with requirements for an input and output voltage. Students should suggest a possible turns ratio which would achieve the required output voltage.

Ask students to turn to the person next to them and explain how a transformer works.

Technician's notes

Be sure to check the latest safety notes on these resources before proceeding.

The following resources are needed for the class demonstration P6d.2:

demonstration transformer
range of mountable coils – for example, 300 turns, 600 turns
digital voltmeters
connecting leads
variable a.c./d.c. power pack
image of transformer computer and data projector

Take care – coils will become hot with use. Minimise the time the power pack is switched on for.

The following resources are needed for the class practical P6d.2a, per group:

insulated wire
C-cores
sticky tape
a.c. power packs (d.c. for extension)
a.c. voltmeters (analogue or digital)

Ensure that students minimise the time their power packs are switched on for. Warn students to take care as wires can get hot during the experiment.

Answers

Page 325

Students' own answers. A shaver socket contains an isolating transformer, which means that there is no direct link between the current that flows through the shaver and the mains supply.

Page 326

1. To increase the strength of the magnetic field.

2. In a step-up transformer, the output voltage is higher than the input voltage (because the secondary coil has more turns than the primary coil). In a step-down transformer, the output voltage is lower than the input voltage (because the secondary coil has less turns than the primary coil).

3. 20 V.

4. 6.0 V.

5. a) 24 V.

   b) 2 A.

   c) 24 A.

P6d.2a Building a transformer

In this practical you will build a transformer.

Apparatus

C-cores

a.c. power supply

insulated wire

connecting leads

crocodile clips

2 × a.c. voltmeters

sticky tape

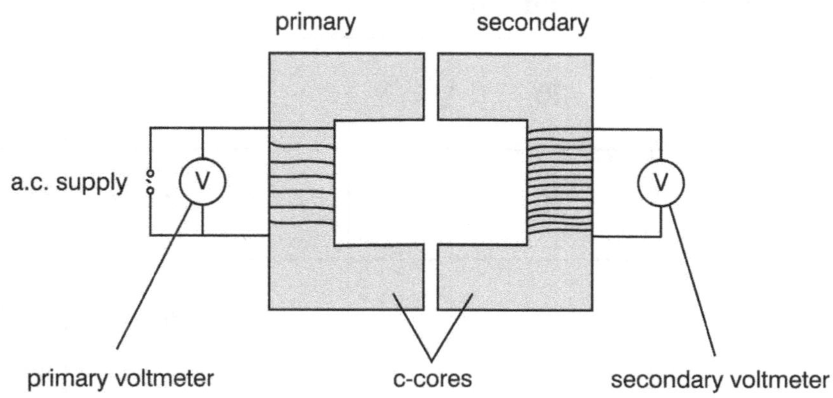

SAFETY INFORMATION
Minimise the time your power pack is switched on for.
Take care as wires can get hot during the experiment.

Method

1. Wind 10 coils of wire onto one of the C-cores. Connect to the power supply and one of the voltmeters.
2. Wind 20 coils of wire onto the other C-core. Connect to a voltmeter.
3. Set the power pack to exactly 4 V and turn on. This is the primary (supply) voltage.
4. Attach the two C-cores together. Take care to ensure they are well aligned.
5. Note the reading on the second voltmeter in the results table below. This is the secondary (output) voltage.
6. Repeat for the remaining results, as follows:

Number of turns on primary	Number of turns on secondary	Primary voltage (V)
20	40	4
40	20	4
40	20	4
20	10	6
20	60	6
30	10	4

Analyse and interpret data

1. Record your results into a table like this:

Number of turns (primary)	Number of turns (secondary)	Turns ratio	Primary voltage (V)	Secondary voltage (V)	voltage ratio
10	20		4		
20	40		4		
40	20		4		
40	20		6		
20	10		6		
20	60		4		
30	10		6		

2. How can the output voltage from a transformer be increased above the input voltage?
3. How can the output voltage from a transformer be decreased below the input voltage?
4. How does the turns ratio relate to the voltage ratio?
5. Explain why a turns ratio greater than 1 produces a step-up transformer.

Activity P6d.3 Using transformers in electricity generation

Specification reference: P6d.6.17P, P6d.6.18P, P6d.6.19P, P6d.6.20P

Learning objectives

- Explain the use of step-up and step-down transformers in the large-scale generation and transmission of electrical energy
- Know and use the relationship: input power = output power
  $V_P I_P = V_S I_S$
  for 100% efficiency

Learning outcomes

- Know how electricity is transmitted and how step-up and step-down transformers are used in the process

Common misconceptions

Work covered on transformers up to this point can give students the impression that a transformer provides 'something for nothing'. Challenge students to weigh up this argument against the law of conservation of energy; gaining electrical energy from nowhere is clearly not possible.

Resources

Student Book pages 326–329

Approach

Introduce the activity

Ask students to list three uses of a step-up transformer and three uses of a step-down transformer.

Challenge students to suggest how a transformer could produce a higher output voltage without increasing either the input voltage or the turns ratio.

Develop the activity

Explain that high voltages are required for the efficient transmission of electrical power but that these same voltages are far too high for use in homes. Therefore, it is necessary to both step-up and step-down the voltage at different points within the system used for transmission of electricity.

Discuss the content of pages 326–327 of the Student Book. It is not immediately obvious to students why power is transmitted at very high voltages. Challenge more able students to consider carefully why this may be the case.

Explain that, while a transformer is able to alter a voltage or current, it is a constant power device – the input power and output power must be equal (assuming 100% efficiency). Remind students that $P = VI$. This should be applied to transformers – hence the equation $V_P \times I_P = V_S \times I_S$. At this level, transformers will be considered to be 100% efficient. The most able could be challenged, however, to consider calculations involving transformers which are less than 100% efficient.

Ask students to work with a partner to tackle the questions on page 329 of the Student Book, which are designed to check understanding.

Bring the activity together

Ask students to devise a transformer calculation for their working partner to attempt. Students should attempt each others' problems and feed back on the quality of the solution offered.

Answers

Page 329

1. Higher voltages mean less energy is wasted in heating the transmission cables, meaning more useful power is delivered at the far end.

2. Transformers change voltages. To send energy at high voltage, firstly the voltage must be increased at the power station end (using a step-up transformer) and then the voltage must be reduced (for safety) at the far end (using a step-down transformer).

3. 40 000 W.

4. 1%.

Activity P6d.4 Consolidation and summary

Learning objectives

- To review the learning points of the topic
- To test understanding through answering questions

Learning outcomes

- Be familiar with the knowledge and understanding summarised in the End of topic checklist
- Be able to apply this knowledge and understanding by answering End of topic questions

Resources

Student Book pages 330–332

Approach

Introduce the activity

Ask students to work with a partner to make a list of key words from this topic. They could then work together to produce a spider diagram showing how the different concepts are linked. They could compare their list with the list of key words given on page 330 of the Student Book. Discuss the checklist given on page 330 and use questioning to see how much of the content students are comfortable with.

Develop the activity

Ask students to work individually to work through the End of topic questions on pages 331–332 of the Student Book without reference to the text. As they work, walk around the classroom observing their answers and questioning them as necessary to find out which questions are causing difficulties.

Bring the activity together

After a set period, ask the students to stop working and discuss any areas of difficulty you observed as you walked round the class. Students should complete any unanswered questions for homework, but you should stress that they should attempt the questions without reference to the text to see how much content they have retained.

Answers

End of topic questions mark scheme

Question	Correct answer	Marks
1	magnetic field lines being cut complete circuit	1 1 1
2	for induction, magnetic field must be changing in direct current, magnetic field is constant so there will be no induction in alternating current, magnetic field is changing so there will be induction	1 1 1 1
3 a)	iron	1
3 b)	a changing magnetic field is induced	1 1 1
3 c)	a changing current is induced	1 1 1
3 d)	$V_P/V_S = n_P/n_S$; $V_P/14 = 12/7$ $V_P = 12/7 \times 14$ = 24 V	1 1 1
4	step-down – the voltage must be reduced	1 1 1
5	the magnetic field cuts the secondary, inducing a voltage	1 1 1 1
6	false – a direct current that is switched on and off rapidly can create the required changing magnetic field in the core	1 1 1
7 a)	step-up	1
7 b)	$V_P = V_S n_P/n_S = 800 \times 150 \div 600$ = 200 V	1 1 1

Question	Correct answer	Marks
8 a)	any turns ratio	1
	1: 1.6	1
	such as 100: 160, 50: 80, 200: 320	1
8 b)	$n_P = n_S V_P / V_S$	1
	= 1600 × 25 000 ÷ 40 000	1
	= 1000	1
9	at points where the voltage is required to be altered, such as increasing voltage	1
	at power station	1
	to transmit along power lines	1
	and decreasing voltage	1
	at substations	1
	so voltage suitable for domestic use	1
10	to reduce I^2R losses	1
	since high voltage	1
	means low current	1
	so squared term	1
	as low as possible	1
	so energy lost in transmission as low as possible	1
11	$I_S = I_P V_P / V_S$ = 20 × 240 ÷ 12	1
	= 400	1
	mA	1
	Total:	52

Section 6: Answers to exam-style questions

Question	Correct answer	Marks
2 a)	suitable lines not touching at least one arrow correctly indicating direction	1 1 1
2 b)	description of using plotting compass OR iron filings	3
2 c) i)	current generates magnetic field around it which interacts with existing field creating a force which causes movement since the coil is free to spin	1 1 1 1
2 c) ii)	increases magnetic field which increases force	1 1
2 c) iii)	reverses direction of magnetic field which reverses direction of force	1 1
	Total:	14

Section 7. Radioactivity, fission and fusion

Contents

Overview of the section

This section has just two topics. The first considers radioactivity. It begins by looking at the structure of an atom in terms of protons, neutrons and electrons. It considers the nature of isotopes. It then goes on to explain the nature and origin of the three types of ionising radiation.

Students learn how to balance nuclear equations. The use of the Geiger–Müller tube to detect ionising radiations is explained. Sources of background radiation are explained. The concept of half-life is introduced and used in calculations of activity. The use of radioactivity in industry and medicine is explained. Radiocarbon dating is discussed. Finally in this section, the dangers of ionising radiations are considered.

The second topic is about particles. The first part of this topic is concerned with nuclear fission. The process is described and then its use in the generation of electricity in nuclear power stations is discussed. The second part of this topic looks at nuclear fusion, compares fusion with fission and then examines the role of fusion as the energy source for stars. It also looks at the possibilities of using nuclear fusion as a source for power generation in the future.

Starting points

The Student Book provides a 'section opener', a double-page spread which sets the scene for students (see pages 336–337). It is structured in this way:

- an overview providing details of the areas of study
- six questions, the first three relating to aspects of physics that should be reasonably familiar to most students, and the final three relating directly to some of the topics to be studied
- a list of section contents.

The questions provide a structure for introducing the section.

1. What is an atom?
2. What happens when charged particles are placed in a magnetic field?
3. What is meant by (a) atomic number, (b) mass number of an element?
4. What do you have to do to balance any equation?
5. What do you understand by the term 'radioactivity'?
6. What happens in a nuclear power station?

The section opener has two main purposes:

- to acknowledge the student's prior learning and to value it
- to provide a benchmark against which future learning can be compared.

The six questions or 'starting points' in the Student Book can be used in a number of different ways to introduce the section:

- You could ask students to consider the questions as an introductory homework task.
- You could put students into groups to share their own ideas and understanding and then report back to the whole class.
- Students could be given access to the Internet, preferably with a tight timescale, to find out the information required.

You could then use a spider chart or other form of wall chart to summarise everybody's ideas. Recording these initial ideas in this way allows you to retain them for reference as the individual topics are developed. In this way, your students' progress in learning can be readily acknowledged.

P7b Radioactivity

Introduction

In this topic, students will learn about the structure of the atom, about the three forms of ionising radiations, how radioactive isotopes decay and are used. The topic ends with a look at the dangers of ionising radiations.

Links to other topics

Section	Essential background knowledge	Useful links
3 Waves		P3c The electromagnetic spectrum
7 Radioactivity, fission and fusion		P7c Fission and fusion

Topic overview

P7b.1	**The structure of an atom** In this activity, students will learn to describe the structure of an atom in terms of protons, neutrons and electrons and use symbols such as ${}^{14}_{6}C$ to describe particular nuclei.
P7b.2	**Ionising radiations** In this activity, students will learn about the three types of ionising radiation, and how to balance nuclear equations.
P7b.3	**Background radiation and half-life** In this activity, students will learn about sources of background radiation. They will then learn about the concept of half-life.
P7b.4	**Uses and dangers of radioactivity** In this activity, students will learn how radioactive isotopes can be useful and about the dangers associated with them.
P7b.5	**Consolidation and summary** This activity provides an opportunity for a quick recap on the ideas encountered in the topic as well as time for the students to answer the End of topic questions in the Student Book.

Activity P7b.1 The structure of an atom

Specification reference: P7b.7.2, P7b.7.3

Learning objectives

- Describe the structure of an atom in terms of protons, neutrons and electrons and use symbols such as ${}^{14}_{6}C$ to describe particular nuclei
- Know the terms atomic (proton) number, mass (nucleon) number and isotope

Learning outcomes

- Be able to describe the structure of an atom in terms of protons, neutrons and electrons
- Be able to use the terms mass (nucleon) number, atomic (proton) number and isotope

Common misconceptions

This area of the Specification deals with objects that we cannot see and for which the experimental evidence is beyond the scope of an International GCSE Physics course. There are a number of terms to learn with similar sounding names and, though the distinction between isotopes is easy to explain in terms of the number of neutrons, talking about the isotopes having the same chemical behaviour but different physical properties is rather advanced for some students.

Students need to realise that the terms 'atomic number' and 'proton number' mean the same, and that the number of protons in an atom equals the number of electrons in the atom.

Resources

Student Book pages 339–341

Worksheet P7b.1a Structure of atoms

Approach

Introduce the activity

Ask students what they understand by the term 'atom'. Generate a list of ideas.

Develop the activity

Discuss the content of pages 340–341 of the Student Book. Make sure that students understand that the atomic number, Z, is the number of the element's place in the periodic table. The chemical properties of an element are determined by the number of protons (and hence the number of electrons).

A different Z number means a different element but a different A (atomic mass number) with the same Z number means an isotope with a different number of neutrons in its nucleus. Explain that radioactive isotopes are decay products of radioactive materials in the Earth when it was formed, elements that have become radioactive as a result of cosmic radiation and also artificial isotopes.

Students should then work in pairs to devise a series of 'what am I?' questions to describe subatomic particles. For example, 'I do not have any electronic charge'.

Students should discuss the questions on page 341 of the Student Book, which are designed to check understanding, before they tackle Worksheet P7b.1a.

Bring the activity together

Ask students to demonstrate their 'what am I?' questions and get the rest of the class to answer them. Then ask students to work in pairs to produce a summary of the content of this activity. After a short period, bring the class back together and check that the summaries cover the stated learning outcomes.

Answers

Worksheet P7b.1a

1.

Element		Subatomic particle		
Name	**Symbol**	**Protons**	**Neutrons**	**Electrons**
carbon	C	6	6	6
cadmium	**Cd**	**48**	64	48
cobalt	Co	27	31	**27**
platinum	**Pt**	78	117	**78**
strontium	Sr	**38**	**50**	**38**
tungsten	**W**	**74**	**110**	**74**
calcium	Ca	**20**	**20**	20

2. a) An ion is an atom that has lost or gained electrons, leaving it with a net charge. Positive ions are formed when an electron in an atom is given enough energy to escape from the atom.

   b) All three isotopes have one proton and so have the same chemical properties. Deuterium has one neutron and tritium has two neutrons. This makes their physical properties different – an obvious difference is the density of the isotopes. Tritium is unstable and decays by beta decay.

   c) Heavy water has a density about 11% greater than water formed with the more common hydrogen atoms – hence the name heavy water.

Answers

Page 341

1. An atom is what all elements are made of.

2. Electrons and protons have equal and opposite charges and there are equal numbers of them in an atom.

3. Atomic number = number of protons in nucleus (and number of electrons in a neutral atom).
Mass number = number of protons + number of neutrons in a nucleus.
Isotopes are nuclei with the same atomic numbers, but different mass numbers.

4. Isotopes of an element have the same number of electrons and it is these that determine the chemical behaviour.

P7b.1a Structure of atoms

1. Particles in atoms

Use a periodic table to help you to complete the following table to show elements and the numbers of particles that make up the atoms of each element. The first has been completed for you.

Element		Subatomic particle		
Name	**Symbol**	**Protons**	**Neutrons**	**Electrons**
carbon	C	6	6	6
			64	48
	Co	27	31	
		78	117	
	Sr			
tungsten				
	Ca			20

2. Ions and isotopes

a) Explain what an ion is and how an ion is created from an electrically neutral atom.
There are three main isotopes of the element hydrogen.
The stable isotopes are hydrogen (^1_1H) and deuterium (^2_1H); the third, tritium, (^3_1H) is unstable with a half-life of around 12 years.

b) Explain how the structure of these three isotopes is similar and how they differ.
Deuterium is sometimes called 'heavy' hydrogen and it combines with oxygen in just the same way as hydrogen to form 'heavy water'.

c) Why is water formed with deuterium called 'heavy water'?

Activity P7b.2 Ionising radiations

Specification reference: P7b.7.4, P7b.7.5, P7b.7.6, P7b.7.7, P7b.7.8

Learning objectives

- Know that alpha (α) particles, beta (β^-) particles and gamma (γ) rays are ionising radiations emitted from unstable nuclei in a random process
- Describe the nature of alpha (α) particles, beta (β^-) particles and gamma (γ) rays and recall that they may be distinguished in terms of penetrating power and ability to ionise
- Describe the effects on the atomic and mass numbers of a nucleus of the emission of each of the four main types of radiation (alpha, beta, gamma and neutron radiation)
- Understand how to balance nuclear equations in terms of mass and charge

Learning outcomes

- Be able to describe the nature and properties of the three major emissions from radioactive materials and know their differing penetrating power and ability to ionise
- Be able to describe what radioactivity is and why some atoms are radioactive

Resources

Student Book pages 343–347

Worksheet P7b.2a Radioactivity

Resources for demonstration (see Technician's notes, following)

Approach

Introduce the activity

The dangers often associated with radiation have been highlighted in the news with the disaster at the Fukushima power plant in Japan (March 2011). The beneficial use of radiation and radioactive materials will be covered in a later activity. Discuss what words students associate with radiation.

Develop the activity

Discuss the content of pages 343–344 of the Student Book. Also discuss the Science in context material about the Large Hadron Collider (LHC).

Demonstrate the penetrating power of different radiations (see Technician's notes below). With the GM tube and counter the following demonstrations highlight the key teaching points:

- The radiation from an alpha-only source is completely stopped by a sheet of paper or 5–10 cm of air.
- The radiation from a beta-only source will pass through paper easily but can be stopped by a relatively thin sheet of low density metal like aluminium.
- Gamma radiation is extremely penetrating and will easily pass through paper and thin aluminium sheet. It is stopped only by several centimetres of lead.

Explain that although gamma radiation is not directly ionising it does produce changes in the structure of atoms (which can lead to the emission of ionising particles) making gamma radiation very dangerous to living organisms. This is discussed in more detail in a later activity. The questions on page 345 of the Student Book provide a good summary.

Now move on to discuss the representation of the various types of ionising radiations, which students will need when they attempt to balance nuclear equations. Alpha, beta, gamma and neutron emission will need to be introduced here. More able students can tackle the Extension feature on page 345, or you may wish to save this for the later activity on uses of radioactivity.

Discuss how to balance nuclear equations using page 347 of the Student Book as a guide. Students can then work in pairs to discuss the questions on page 347. They can then tackle Worksheet P7b.2a as individuals.

Bring the activity together

Ask students to work with a partner to devise a quiz about radioactive particles or write instructions for balancing nuclear equations. After a short while, ask students to give you some examples of questions from their quiz for other students to answer, or read out their instructions and see whether other students can follow them.

Technician's notes

Be sure to check the latest safety notes on these resources before proceeding.

The following resources are needed for the class demonstration P7b.2:

Geiger–Müller tube or solid state detector
counter/rate meter
radioactivity kit including a selection of low-level alpha, beta and gamma sources, and various thicknesses of lead and aluminium sheet

These must be kept securely and handled with appropriate tongs following safe practice.

Answers

Worksheet P7b.2a

1. a) Alpha and beta.
   b) Gamma.
   c) Very energetic (high frequency) electromagnetic waves.
   d)

Name	Notation
alpha	${}^{4}_{2}He$
beta	${}^{0}_{-1}e$

2. a) The tracks appear suddenly and unpredictably with different interval between them and different directions.
   b) The tracks are all about the same length so the alpha particles are losing their energy after roughly the same number of collisions with air molecules – so they must have similar energies.
   c) Students' own table.

3. a) $^{239}_{92}\mathrm{U} \rightarrow {}^{239}_{94}\mathrm{Np} + {}^{0}_{-1}\boxed{\mathrm{e}}$ Decay of uranium by **beta** emission.

b) $^{185}_{79}\mathrm{Au} \rightarrow {}^{181}_{77}\mathrm{Ir} + {}^{4}_{2}\boxed{\mathrm{He}}$ Decay of gold by **alpha** emission.

c) $^{24}_{11}\mathrm{Na} \rightarrow {}^{24}_{12}\mathrm{Mg} + {}^{0}_{-1}\mathrm{e}$ Decay of sodium by **beta** emission.

d) $^{4}_{2}\mathrm{He} \rightarrow {}^{6}_{3}\mathrm{Li} + {}^{0}_{-1}\boxed{\mathrm{e}}$ Decay of helium by **beta** emission.

e) $^{232}_{90}\mathrm{Th} \rightarrow {}^{228}_{88}\mathrm{Ra} + {}^{4}_{2}\boxed{\mathrm{He}}$ Decay of thorium by **alpha** emission.

Page 345 (top)

1. The emission of particles and/or energy from an unstable nucleus.

2. Unstable.

3. Alpha particles are relatively large, so as they travel they collide often, reducing their energy quickly, so they cannot travel very far. Beta particles are smaller and travel more quickly, having fewer collisions and losing less energy each time. This means they will travel further before losing all their energy. Gamma radiation, being electromagnetic waves, only interacts weakly with matter so it has a much larger range.

Page 345 (bottom)

Students' own answer should include the following:

Amount of beta radiation picked up by detector is very sensitive to thickness of the paper and so accurate.

Beta radiation is absorbed by a thin sheet of aluminium, so the source can be shielded easily.

Page 347

1. Two protons and two neutrons are emitted. Atomic number reduced by 2 and mass number by 4.

2. A neutron changes into a proton. Atomic number increases by 1. Mass number not affected.

3. No change to the number or types of particles. There is a reduction in energy however.

4. Emitting alpha or beta radiation changes the number of protons in the nucleus. This means the number of outer electrons will also change, changing the chemical behaviour – it is a different element.

P7b.2a Radioactivity

1. Types of radiation

You have learnt that two types of nuclear radiation are **particles** which may be ejected from the nucleus of an unstable isotope of an element.

a) Name these particles.
The third type of nuclear radiation does not consist of particles.

b) Name this type of radiation.

c) What does this type of radiation consist of?

As these particles have mass and charge they may be represented using similar notation to that used for the elements themselves (^{A}X), where A is the mass number and Z is the number of **positive** charges that the particle possesses.

d) Complete this table showing how they are represented in this way.

Name	Notation
	$^{?}_{?}\mathrm{He}$
	$^{?}_{?}\mathrm{e}$

2. Radiation observations

a) The tracks observed in the cloud chamber demonstration help us to understand that radioactive decay is random. Explain how the observations that you have made do this.

b) As the alpha radiation travels through the air and alcohol vapour it ionises the air and the alcohol vapour condenses on the ions.

Each time the radiation ionises an air molecule it loses some of its energy. All the alpha emissions have roughly the same amount of energy.

Explain the feature of the cloud chamber tracks that shows this.

c) Summarise the properties of α, $\beta^{-}$and γradiation in a table. You should consider the nature of each type of radiation, its charge, its range in air and its ionising effect.

3. Balancing equations

Complete the following nuclear decay equations by filling in the empty boxes and stating the type of decay process.

$^{239}_{92}\mathrm{U} \rightarrow {}^{239}_{94}\mathrm{Np} + {}^{?}_{-1}\boxed{?}$ Decay of uranium by .... emission

$^{185}_{79}\mathrm{Au} \rightarrow {}^{?}_{77}\mathrm{Ir} + {}^{4}_{?}\boxed{?}$ Decay of gold by .... emission

$^{24}_{?}\mathrm{Na} \rightarrow {}^{24}_{12}\mathrm{Mg} + {}^{?}_{?}\mathrm{e}$ Decay of sodium by .... emission

$^{?}_{2}\mathrm{He} \rightarrow {}^{6}_{3}\mathrm{Li} + {}^{0}_{?}\boxed{?}$ Decay of helium by .... emission

$^{?}_{?}\mathrm{Th} \rightarrow {}^{228}_{88}\mathrm{Ra} + {}^{4}_{2}\boxed{?}$ Decay of thorium by .... emission

Activity P7b.3 Background radiation and half-life

Specification reference: P7b.7.9, P7b.7.10, P7b.7.11, P7b.7.12, P7b.7.13

Learning objectives

- Know that ionising radiations can be detected using a photographic film or a Geiger–Müller detector
- Explain the sources of background (ionising) radiation from Earth and space
- Know that the activity of a radioactive source decreases over a period of time and is measured in becquerels
- Know the definition of the term half-life and understand that it is different for different radioactive isotopes
- Use the concept of half-life to carry out simple calculations on activity, including graphical methods

Learning outcomes

- Be able to describe how radioactivity can be detected and is measured
- Be able to use ideas about half-life to describe how radioactive materials behave

Resources

Student Book pages 348–352

Resources for class demonstration (see Technician's notes, following)

Approach

Introduce the activity

Cosmic radiation, including that from the Sun, can be harmful to humans. Discuss how the Earth's atmosphere provides protection from this type of radiation.

Develop the activity

Discuss the content of page 348 of the Student Book. Use a Geiger counter to demonstrate that background radiation is present all the time. Explain that ionising radiations can also be detected using photographic film. More able students can tackle the extension feature on page 348 which considers radiation doses for different sources.

The Developing investigative skills feature provides further material for students on this topic.

Now discuss the content of pages 349–350 of the Student Book which deals with how activity decreases over time, and the concept of half-life.

Demonstrate the half-life concept with a coin tossing experiment. Students start by standing with either hands on heads, or behind backs to signify heads or tails. A coin is flipped and those that chose incorrectly sit down. This is repeated several times – results could be plotted to show exponential decay. This works best with a larger group. The speed of coin flipping will affect the 'half-life'.

Now allow students to tackle the worked example on page 350 without the given solution. After a short period, bring the class back together and discuss students' solutions and how they compare with the model answer.

Next, students should work on the Developing investigative skills feature on pages 351–352 with a partner. After a period of time, bring the class back together and compare student responses to the model answers.

Finally, allow students to discuss the questions on page 352 of the Student Book, which are designed to check understanding.

Bring the activity together

Discuss whether low dosage background radiation has an impact on our health. Set the students some problems on half-life.

Technician's notes

Be sure to check the latest safety notes on these resources before proceeding.

The following resources are needed for the class demonstration P7b.3:

Geiger counter
coin

Answers

Page 348

1. Cosmic rays, radiation from rocks, radon gas, radioactive isotopes in the body.
2. Medical sources, consumer products and others.
3. Medical sources.
4. X-rays, gamma rays (from radiotherapy).
5. Nuclear power stations, atomic bombs.

Developing investigative skills, pages 349

1. Geiger–Müller tube

2. Use tongs or forceps to hold the sources, shielded sources, keep distance from the sources, do not point the sources at living tissues, spend as little time near the sources as possible.

3. U is unaffected by paper or aluminium. It is only absorbed by lead so it is a gamma source.

V reduces sharply with all absorbers. Paper absorbs the radiation which must be alpha.

W is partly stopped by paper so there is some alpha present. Some more radiation is also stopped by aluminium so there must be beta radiation too.

X is partly stopped by paper so alpha is present. The remainder of the radiation is unaffected by aluminium but stopped by the lead so it must be gamma.

4. There is only one more count with paper (202 without and 203 with). This is a very small % difference and within the limits of the errors that can be expected. Also, radioactive decay is a random process which easily accounts for this small fluctuation.

Developing investigative skills, pages 351–352

1. The ionising radiation from radioactive sources can be very harmful and the use of radioactive materials is covered by a range of regulations. Young people are generally not allowed to handle radioactive materials, particularly as young people are still growing and developing rapidly. Reducing the exposure to radioactive materials reduces the risk of any harmful effects to cells in the body.

2. With no radioactive source present, the teacher should take a count rate several times to find a mean background activity. This value should be subtracted from the values measured during the experiment, since this count would have been expected anyway and cannot be due to the protactinium sample.

3. The independent variable was the time since the experiment started, the dependent variable the activity of the sample.

4. Graph is a smooth curve.

5. Depending on the particular curve drawn, the half-life should be a little over one minute (actual value for this isotope is 72 s). There should be evidence on the graph of how the value was found – e.g. lines traced across from one axis to the other in two places, where one has half the activity of the other.

6. Random in this context indicates that the behaviour of any individual atom cannot be predicted. Towards the end of the experiment, the values do not follow the pattern exactly, indicating that the randomness of the decay is becoming more significant.

7. The experiment is to investigate the behaviour of the protactinium sample. Background radiation is not connected to the sample, but it will change the measurements made, possibly misleading us to draw incorrect conclusions about the behaviour of protactinium.

8. To get an accurate value for the half-life we need a graph with as smooth a curve as possible – one where we are confident that we have drawn the correct line. The larger the sample of the radioactive source, the higher values our measurements will have and the effect of random variations will be reduced. If we repeat the test and **add all the values together** this would effectively give us a larger sample of the material and should lead to a more accurate value for the half-life. However, simply repeating the experiment and averaging the values of half-life obtained would not significantly improve the accuracy.

Page 352

1. Radioactivity in soil, rocks and materials like concrete, radioactive gases in the atmosphere and cosmic rays from outer space.

2. Half-life is the time taken for half of the unstable nuclei in a sample to decay: that is, the time for the activity to reduce to half its current level.

3. 2 hours.

4. 4 hours.

5. Activity is 800 Bq now, so in 8 hours it will be 400 Bq. 8 hours later it will be 200 Bq and a further 8 hours later the activity will be 100 Bq. This gives a total of 24 hours, three half-lives.

Activity P7b.4 Uses and dangers of radioactivity

Specification reference: P7b.7.14, P7b.7.15, P7b.7.16

Learning objectives

- Describe the uses of radioactivity in industry and medicine
- Describe the difference between contamination and irradiation
- Describe the dangers of ionising radiations, including: radiation can cause mutations in living organisms, radiation can damage cells and tissue, the problems arising in the disposal of radioactive waste.
- Describe how the risks associated with ionising radiations can be reduced

Learning outcomes

- Be able to describe some uses and hazards associated with radioactive materials

Resources

Student Book pages 352–356

Approach

Introduce the activity

Hold a discussion to see what students already know (or think they know) about the different applications for ionising radiation, e.g. medicine, power generation, food treatment.

Develop the activity

Discuss the content on pages 352–354 of the Student Book about uses of radioactivity. You could divide the class into small groups to carry out further research on the different uses. Some students may need more guidance in their research. The pages in Wikipedia would be a good starting point.

Discourage them from copying and pasting of material which is beyond their level of understanding and which is not relevant to the Specification. The emphasis should be on why a particular isotope is used for a given purpose.

Stress that radioisotopes have the same chemistry as non-radioactive isotopes of the same element, and thus are treated, for example, by the human body, in exactly the same way.

Now discuss the dangers of ionising radiation using page 354 of the Student Book for guidance. Explain the difference between contamination and irradiation and discuss the problem of radioactive waste, using the text on page 355 of the Student Book. Finally, allow students to work in pairs to consider the questions on page 355 which are designed to check understanding. The Science in context material about dating the Earth (pages 355 and 356 in the Student Book) can also be looked at.

Bring the activity together

As a class, produce a spidergram summarising the main content of this activity.

Answers

Page 355

1. Sterilising medical equipment and preserving food.

2. A radioactive substance with a half-life long enough for it to be spread out and be detected. Medical tracers are used to detect blockages in vital organs. Agricultural tracers monitor the flow of nutrients through a plant. Industrial tracers can measure the flow of liquid and gases through pipes to identify leakages.

3. For example: medical uses – gamma tracers, radioactive iodine to target the thyroid gland; non-medical – dating of rocks, smoke detectors.

4. Radioactive emissions are ionising radiations – they can ionise cells in the body which may destroy the cells or damage them (particularly hazardous is the damage that involves mutations to the cell which can lead to cancerous changes).

5. Irradiation is when a material is exposed to alpha, beta or gamma radiation but it does not become radioactive itself. Contamination occurs when an organism ingests a radioactive material. This can sometimes stay in the body for many years. Other materials can also be contaminated when they have absorbed or are coated with radioactive dust, for instance.

Activity P7b.5 Consolidation and summary

Learning objectives

- To review the learning points of the topic
- To test understanding through answering questions

Learning outcomes

- Be familiar with the knowledge and understanding summarised in the End of topic checklist
- Be able to apply this knowledge and understanding by answering End of topic questions

Resources

Student Book pages 357–361

Approach

Introduce the activity

Ask students to work with a partner to make a list of key words from this topic. They could then work together to produce a spider diagram showing how the different concepts are linked. They could compare their list with the list of key words given on page 357 of the Student Book. Discuss the checklist given on page 357 and use questioning to see how much of the content students are comfortable with.

Develop the activity

Ask students to work individually to work through the End of topic questions on pages 358–361 of the Student Book without reference to the text. As they work, walk around the classroom observing their answers and questioning them as necessary to find out which questions are causing difficulties.

Bring the activity together

After a set period, ask the students to stop working and discuss any areas of difficulty you observed as you walked round the class. Students should complete any unanswered questions for homework, but you should stress that they should attempt the questions without reference to the text to see how much content they have retained.

Answers

End of topic questions mark scheme

<table>
<tr><th>Question</th><th>Correct answer</th><th>Marks</th></tr>
<tr><td>1</td><td>Completed table as shown
<table>
<tr><th>Atom</th><th>Symbol</th><th>Number of protons</th><th>Number of neutrons</th><th>Number of electrons</th></tr>
<tr><td>hydrogen</td><td>$^{1}_{1}H$</td><td>1</td><td>0</td><td>1</td></tr>
<tr><td>carbon</td><td>$^{12}_{6}C$</td><td>6</td><td>6</td><td>6</td></tr>
<tr><td>calcium</td><td>$^{40}_{20}Ca$</td><td>20</td><td>20</td><td>20</td></tr>
<tr><td>uranium</td><td>$^{238}_{92}U$</td><td>92</td><td>146</td><td>92</td></tr>
</table>
</td><td>12</td></tr>
<tr><td>2 a)</td><td>11 protons
13
neutrons</td><td>1
1
1</td></tr>
<tr><td>2 b)</td><td>the count falls from 100 Bq to 50 Bq in 15 hours. It also falls from 50 Bq to 25 Bq in
15
hours</td><td>1
1
1</td></tr>
<tr><td>3 a)</td><td>beta particle</td><td>1</td></tr>
<tr><td>3 b)</td><td>$x = 24$
$y = 12$</td><td>1
1</td></tr>
<tr><td>4</td><td>isotopes of an element all have the same atomic number
but have different atomic masses because they have different numbers of neutrons in their nuclei</td><td>1
1</td></tr>
<tr><td>5</td><td>beta particles are high energy (i.e. fast moving) electrons;
they are negatively charged
and have very little mass</td><td>1
1
1</td></tr>
</table>

Question	Correct answer	Marks
6	ionisation means knocking electrons away from an atom leaving it with a net positive charge (it becomes a positive ion)	1 1
7	$^{220}_{86}\text{Rn} \rightarrow\ ^{216}_{84}\text{Po} +\ ^{4}_{2}\text{He}$	3 (1 mark for each correct nuclide)
8	$^{14}_{6}\text{C} \rightarrow\ ^{14}_{7}\text{N} +\ ^{0}_{-1}\text{e}$	3 (1 mark for each correct nuclide)
9	place each source in turn close (no more than a few cm) to a GM tube connected to a counter place a sheet of paper between each source and the detector for the alpha source only the count rate would fall to the background level as alpha particles are blocked by paper with the remaining two sources repeat the experiment using a sheet of aluminium about 1–2 mm thick if the count falls to the background level you have a beta source as 1–2mm of aluminium will block beta particles but has very little effect on gamma rays	1 1 1 1 1 1
10	in experimental work in schools the activity of materials being investigated is, for health and safety reasons, low neglecting to exclude background radiation from your measurements would have a significant effect on their accuracy to do this you would check background radiation level using Geiger counter	1 1 1
11 a)	200 it will have halved twice	1 1
11 b)	4	2

Question	Correct answer	Marks
11 c)	3 half-lives so 1 hour 30 minutes	1 1
12 a)	16 counts/min	1 1
12 b)	40 days	1 1
13	8 minutes	2 1
14 a)	about 8 days	1
14 b)	about 48 days	1
14 c)	about 56 days	1
14 d)	about 72 days	1
15	set up experiment with source A pointing at a radiation detector at a fixed distance away record the counts detected in 1 minute repeat the experiment with a paper barrier in between the source and the detector and again record the counts in 1 minute if the count rate falls significantly it is an alpha source similarly using in turn an aluminium barrier and then a lead barrier will reveal whether it is a beta source (reduced by a few mm of aluminium) or a gamma source (reduced by a cm of lead) repeat this procedure for sources B and C	1 1 1 1 1 1 1 1
16 a)	a tracer is a radioactive isotope used in detection: tracers are widely used to detect leaks and blockages	1
16 b)	sodium-24 the lawrencium-257 has too short a half-life the sulfur-35 and carbon-14 have half-lives which are too long	1 1 1 1

Question	Correct answer	Marks
17 a)	gamma penetrates aluminium easily this means the count rate will not change with thickness alpha radiation will not penetrate aluminium this means the count rate will not change with thickness	1 1 1 1
17 b)	beta radiation count is measured when the foil is at the required thickness (this is called calibration) when the manufactured foil is moving through the beta its count rate will vary depending on the thickness: too thick and the count rate is reduced too thin and the count rate is too high a message can be sent to the thickness control unit to adjust the thickness	1 1 1 1
17 c)	the count rate will reduce over time the machine will need recalibrating otherwise it will deduce that the aluminium is thinner than it is so the thickness will be incorrect	1 1 1 1
	Total:	74

P7c Fission and fusion

Introduction

In this topic, students will learn about nuclear fission and nuclear fusion.

Links to other topics

Section	Essential background knowledge	Useful links
7 Radioactivity, fission and fusion	P7b Radioactivity	

Topic overview

P7c.1	**Nuclear fission** In this activity students learn about the process of nuclear fission.
P7c.2	**Nuclear fusion** In this activity students learn about the process of nuclear fusion.
P7c.3	**Consolidation and summary** This activity provides an opportunity for a quick recap on the ideas encountered in the topic as well as time for the students to answer the End of topic questions in the Student Book.

Activity P7c.1 Nuclear fission

Specification reference: P7c.7.17, P7c.7.18, P7c.7.19, P7c.7.20, P7c.7.21, P7c.7.22

Learning objectives

- Know that nuclear reactions, including fission, fusion and radioactive decay, can be a source of energy
- Understand how a nucleus of U-235 can be split (the process of fission) by collision with a neutron, and that this process releases energy as kinetic energy of the fission products
- Know that the fission of U-235 produces two daughter nuclei and a small number of neutrons
- Describe how a chain reaction can be set up if the neutrons produced by one fission strike other U-235 nuclei
- Describe the role played by the control rods and moderator when the fission process is used as an energy source to generate electricity
- Understand the role of shielding around a nuclear reactor

Learning outcomes

- Be able to describe how our knowledge of the atom has led to nuclear fission as an energy resource

Common misconceptions

This is very much a topical issue and students should be encouraged to understand the benefits and hazards of using nuclear power.

Resources

Student Book pages 362–366

Worksheet P7c.1a Nuclear fission

Resources for class demonstration: a set of dominoes.

Approach

Introduce the activity

Remind students about what they know about the model of the atom we use today. (This is covered in Activity P7b.1 'The structure of an atom'.) Ask them to tell you what particles are found in an atom.

Develop the activity

Discuss the content of page 363 of the Student Book. Take time to ensure that students understand that the fission of U-235 produces daughter nuclei and neutrons, which can go on to split further U-235 atoms in a chain reaction.

Set up a row of dominoes to demonstrate a chain reaction as described on page 364 of the Student Book. Ask students to work with a partner to discuss the questions on page 354 to check understanding.

Now move on to consider how nuclear fission is used in a nuclear reactor. Make sure that students appreciate that this is a controlled nuclear reaction, and that they appreciate the role of the control rods. Use pages 365–366 of the Student Book as guidance.

More able students can tackle the Extension feature on page 366 of the Student Book. Students should work with a partner to tackle the questions on page 366 to check understanding.

Bring the activity together

Ask students to work with a partner to produce a summary of this activity. After a while, bring the class back together to check that the main learning outcomes listed above have been picked up.

Answers

Worksheet P7c.1a

1. See diagram on right.
2. a) A (slow) neutron.
   b) 2 or 3.
   c) More than 1.
   d) 1.
   e) Critical mass.
   f) Control rods (usually made from boron) absorb neutrons to stop them going on to cause new fission reactions. These rods can be raised out of the nuclear core to cause the reaction to speed up or fully inserted into the core to shut down the nuclear reaction completely.

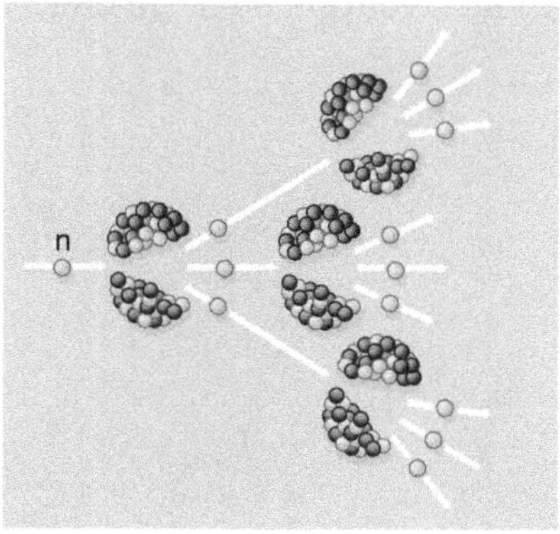

Page 364

1. The process where a large nucleus absorbs a neutron and then splits into two large fragments with the release of energy and further neutrons.

2. A lot of energy is given out, the nucleus splits into two parts and some neutrons are emitted.

3. The neutrons released during a fission can collide with further nuclei, causing further fissions which release further neutrons for further fissions, and so on.

Page 366 (top)

1. Fuel rods – where the fission reaction happens releasing energy as thermal energy. Control rods – act as 'sponges' absorbing neutrons to control the rate of fission.

Moderator – slows the neutrons down so they are at the correct speed to be absorbed for further fission reactions.

2. By circulating pressurised water or another fluid through the reactor.

3. The waste is radioactive, some of it highly radioactive, so it needs to be handled, transported and stored carefully to reduce the risks to people.

Page 366 (bottom)

1. The control rods absorb some of the neutrons, which prevents them going on to split further nuclei. The number of nuclei that undergo fission is thus controlled and so the energy released is also controlled.

2. Take the control rods out of the core/further out of the core.

3. Move the control rods further into the core.

P7c.1a Nuclear fission

1. Nuclear fission

The diagram shows the start of a nuclear fission reaction. Copy and complete the diagram to show how the three neutrons each go on to split a further atom.

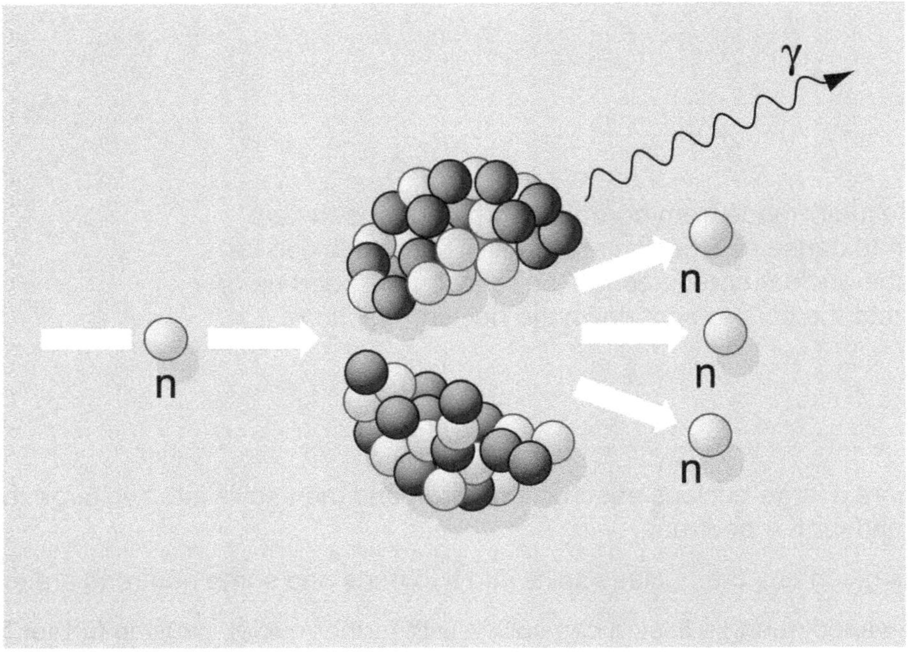

2. Chain reactions

a) What type of particle is used to trigger a nuclear fission in an atom of nuclear fuel?

b) Roughly how many of these particles are produced by the fission of an atom of nuclear fuel?

c) For a chain reaction to take place, how many of these particles must go on to cause fission in more atoms of nuclear fuel?

d) How many of these particles must go on to cause fission in more atoms of nuclear fuel for the reaction to just be a self-sustaining reaction?

e) If the amount of nuclear fuel is too small, the neutrons mass will not go on to trigger further fissions and the reaction will stop. There is a minimum amount of nuclear fuel needed for the reaction to continue – what is this amount called?

f) How is the rate of the reaction controlled in a nuclear reactor?

Activity P7c.2 Nuclear fusion

Specification reference: P7c.7.23, P7c.7.24, P7c.7.25, P7c.7.26

Learning objectives

- Explain the difference between nuclear fusion and nuclear fission
- Describe nuclear fusion as the creation of larger nuclei resulting in a loss of mass from smaller nuclei, accompanied by a release of energy
- Know that fusion is the energy source for stars
- Explain why nuclear fusion does not happen at low temperatures and pressures, due to electrostatic repulsion of protons

Learning outcomes

- Be able to describe the nature of attractive and repulsive forces inside the nucleus
- Be able to explain how energy is absorbed and released in a fusion reaction
- Know the elements of nuclear fusion in stars
- Appreciate the longevity of nuclear fusion in stars

Resources

Student Book pages 367–368

Access to Internet and presentation facilities such as a Smartboard

Approach

Introduce the activity

The students will need to be made aware of the basic ideas of fusion. Those being:

- the nature of the forces inside the nucleus
- the energy needed to get nuclei close enough to each other
- small nuclei, when undergoing fusion absorb and release energy
- fusion happens in stars

Develop the activity

The students can then research nuclear fusion in stars to study the mechanisms involved in fusion.

In this activity they could research the following areas:

- Which nuclei (of which elements) undergo fusion in stars?
- What is the fusion reaction in stars?
- Where does the energy come from to undergo fusion in stars?
- How long do these fusion reactions last in stars?
- How much energy is released in fusion?

Bring the activity together

They could present their findings as a power point which could be shared with others in the class. The class can then contribute their ideas and add to the learning.

The basic ideas can then be built on and summarised to the students:

- Inside a nucleus there are repulsive forces between the positive protons. On their own they would cause the nucleus to fall apart. The only thing that holds the nucleus together is the strong nuclear forces between the particles in the nucleus.
- For nuclear fusion to happen, the nuclei must be close together so that the attractive nuclear force between the protons and neutrons overcomes the repulsive electric force between the protons. Nuclear forces only work over small distances so it needs lots of energy to bring the nuclei together. That is why fusion mainly happens in high temperature surroundings, where the particles are very close together and moving very quickly. When small nuclei such as hydrogen and helium are fused together, more energy is released than is absorbed.
- On Earth we depend on nuclear fusion for heat and light. Without it we would have no weather, plants, life, etc.

The activity can be further developed along the lines of nuclear fusion reactions on Earth used to provide power generation. The energy produced in fusion is vast and scientists have wondered (or dreamed) of having fusion reactions on Earth that could be used to provide clean and cheap energy in the future. Some scientists have claimed that they successfully produced 'cold fusion' reactions.

The students can research these ideas perhaps to cover these points:

- fusion reaction – has someone really shown it can be done on Earth?
- What are the practical difficulties of nuclear fusion for energy production?

Here on Earth, scientists have only been able to achieve fusion reactions lasting a few seconds but around the world there has been research into using nuclear fusion as an energy source. If successful, nuclear fusion could possibly result in a limitless supply of clean energy.

Activity P7c.3 Consolidation and summary

Learning objectives

- To review the learning points of the topic
- To test understanding through answering questions

Learning outcomes

- Be familiar with the knowledge and understanding summarised in the End of topic checklist
- Be able to apply this knowledge and understanding by answering End of topic questions

Resources

Student Book pages 369–371

Approach

Introduce the activity

Ask students to work with a partner to make a list of key words from this topic. They could then work together to produce a spider diagram showing how the different concepts are linked. They could compare their list with the list of key words given on page 369 of the Student Book. Discuss the checklist given on page 369 and use questioning to see how much of the content students are comfortable with.

Develop the activity

Ask students to work individually to work through the End of topic questions on pages 370–371 of the Student Book without reference to the text. As they work, walk around the classroom observing their answers and questioning them as necessary to find out which questions are causing difficulties.

Bring the activity together

After a set period, ask the students to stop working and discuss any areas of difficulty you observed as you walked round the class. Students should complete any unanswered questions for homework, but you should stress that they should attempt the questions without reference to the text to see how much content they have retained.

Answers

End of topic questions mark scheme

Question	Correct answer	Marks
1	alpha radiation can be stopped by a few sheets of paper beta by a few mm of aluminium and gamma by thick lead neutron radiation can be stopped by very thick lead, steel, concrete or water power stations therefore need to use thick layers of lead, steel, concrete and sometimes water.	1 1 1 1 1
2	nuclear fission is when large nuclei split to form smaller nuclei fusion is when small nuclei are joined together to produce a larger nuclei	1 1
3	in nuclear fission, an unstable atom is triggered into splitting apart by absorbing a neutron	1 1 1
4	the fission of uranium-235 produces two lighter nuclei plus three neutrons each of these neutrons can trigger another atom to split	1 1 1
5	the reaction proceeds with increasing speed as more and more neutrons are produced by fission, setting off further fissions at the same time the amount of energy produced as heat is rapidly increasing rapidly making the task of controlling the chain reaction impossible	1 1 1
6	the massive and rapid release of huge amounts of energy	1 1 1
7	by controlling the rate of the fission reaction so that it proceeds at a self-sustaining but much slower rate this is done using control rods to absorb neutrons so the rapid escalation cannot occur	1 1 1

Question	Correct answer	Marks
8	as a nucleus splits it releases energy because the energy of the nucleus before the fission is greater than the combined energy of the two daughter nuclei in a chain reaction this amount of energy is multiplied many times and so a great deal of energy is released in a chain reaction	1 1 1 1 1 1
9	in a nuclear power station, neutrons collide with U-235 nuclei which leads to fission of the nuclei some of the neutrons produced in the fission reaction are absorbed by control rods to prevent the reaction from getting out of control	1 1 1 1 1 1
10	nuclear fusion requires high temperatures and high pressures to bring the nuclei close enough together these conditions are difficult to manage on Earth in a safe and controlled way also, currently the reactions require more energy than they produce	1 1 1 1 1
11 a) i)	no evidence just a statement: 'add to that impact on the local environment'	1 1
11 a) ii)	company spokesperson says emissions are water vapour and steam which is cleaner than emissions from coal-fired	1 1
11 a) iii)	concerns over disposal of waste fear of the unknown	1 1
11 b)	students' own opinions – give credit for well-balanced views supported by scientific knowledge	6
11 c)	students' own research	6
	Total:	56

Section 7: Answers to exam-style questions

Question	Correct answer	Marks
2 a)	(nuclear) fission	1
2 b)	kinetic energy of the particles made	1 1
2 c)	neutrons released can collide with further nuclei causing further fissions	1 1 1
2 d) i)	absorb neutrons controlling rate of fission	1 1
2 d) ii)	slow down neutrons allowing them to be absorbed	1 1
	Total:	10

Contents

Overview of the section

The Motion in the Universe topic is a brief introduction to astronomy. It covers gravitational field strength, gravitational force, and the orbits of comets, moons and planets. It also explores the how the Universe, galaxies and our Solar System are related.

Stellar evolution explores the classification of stars and how this is related to their temperature and colour. Absolute and apparent magnitude is explained and the Hertzsprung–Russell diagram (HR diagram) is introduced. The life cycles and stages for stars of different masses are also compared.

The origins of the Universe are considered in the Cosmology topic where the Big Bang and observed evidence is explained. The Doppler effect is revisited and applied to the observation of red-shift which is explained both qualitatively and quantitatively.

Starting points

The Student Book provides a 'section opener', a double-page spread which sets the scene for students (see pages 374–375). It is structured in this way:

- an overview providing details of the areas of study
- four questions: the first about the Solar System will have partly been addressed in pre-GCSE studies; the contents of the Solar System are often familiar, although a clear understanding of the differences between galaxies, stars, moons and planets can often be lacking; the 'Big Bang' as a name may be familiar to many but this section aims to increase understanding in this area; finally, the details of the life cycle of stars will be new to many and it will help to relate the behaviour of stars to the ideas of nuclear fusion studied in Section 7.
- a list of section contents.

The questions provide a structure for introducing the section.

1. What is the Solar System made of?
2. What are the differences between galaxies and stars, a red giant and a white dwarf?
3. What is the Universe and how may it have started?
4. What is meant by the life cycle of a star?

The section opener has two main purposes:

- to acknowledge the student's prior learning and to value it
- to provide a benchmark against which future learning can be compared.

The four questions or ‘starting points’ in the Student Book can be used in a number of different ways to introduce the section:

- You could ask students to consider the questions as an introductory homework task.
- You could put students into groups to share their own ideas and understanding and then report back to the whole class.
- Students could be given access to the Internet, preferably with a tight timescale, to find out the information required.
- You could then use a spider chart or other form of wall chart to summarise everybody’s ideas. Recording these initial ideas in this way allows you to retain them for reference as the individual topics are developed. In this way, your students’ progress in learning can be readily acknowledged.

P8b Motion in the Universe

Introduction

In this topic students are introduced to the basics of astronomy. They learn about the gravitational force and gravitational field strength. They learn that the Universe is a large collection of billions of galaxies, that a galaxy is a large collection of billions of stars and that our Solar System is in the Milky Way galaxy.

Links to other topics

Section	Essential background knowledge	Useful links
1 Forces and motion	P1b Movement and position P1c Forces, movement, shape and momentum	http://www.bbc.co.uk/schools/gcsebitesize/science/ocr_gateway_pre_2011/living_future/6_solar_system1.shtml

Topic overview

P8b.1	**The Solar System** In this activity students learn to name and describe objects in the Solar System – planets, comets, moons and satellites.
P8b.2	**Gravitational field strength** In this activity students learn about gravitational field strength. The activity concludes by moving out beyond the Solar System to other galaxies and the entire Universe.
P8b.3	**Consolidation and summary** This activity provides an opportunity for a quick recap on the ideas encountered in the topic as well as time for the students to answer the End of topic questions in the Student Book.

Activity P8b.1 The Solar System

Specification reference: P8b.8.2, P8b.8.5, P8b.8.6

Learning objectives

- Know that the Universe is a large collection of billions of galaxies, that a galaxy is a large collection of billions of stars and that our Solar System is in the Milky Way galaxy
- Describe the differences in the orbits of comets, moons and planets
- Use the relationship between orbital speed, orbital radius and time period:

$$\text{orbital speed} = \frac{2 \times \pi \times \text{orbital radius}}{\text{time period}}$$

$$v = \frac{2 \times \pi \times r}{T}$$

Learning outcomes

- Be able to distinguish between planets, moons, comets, artificial satellites and stars
- Be able to sketch the different orbits of these objects
- Be able to calculate the orbital speed of an object in orbit

Common misconceptions

Widespread light pollution means that many students will never have seen the night sky clearly as our ancestors did, so they may not appreciate how compelling it is to find out what is there.

Resources

Student Book pages 377–382

Resources for the introduction to the activity: Pictures of objects in the Universe – these can be obtained for free from the NASA website.

Approach

Introduce the activity

Create a spider diagram or concept map to find out what your students think is in space. Then ask them to say where the scientific evidence comes from.

Manned spaceships have visited and landed on our Moon from 1969. Other places have had unmanned spaceships, or probes, visit them. Some of these probes have taken measurements or photographs. Others have been deliberately crash landed onto the surface of other objects to give feedback to scientists on Earth. Amongst these landing sites are the Moon, Mars, Venus, Titan (a moon of Saturn) a few asteroids (impact on Itokawa, which is an asteroid) and a comet. Comet 67P/Churyumov–Gerasimenko was visited by the Philae lander, which had a successful soft landing. Other evidence mostly comes from the electromagnetic radiation which the objects emit. A useful STEM resource 5 minute videoclip covers a summary from NASA about the Viking landers: https://www.stem.org.uk/elibrary/collection/2955. You could also run a picture show of objects in the Universe using images from the NASA website.

Develop the activity

Discuss 'The Solar System' on page 378 of the Student Book. Explain that one astronomical unit is the mean distance of the Earth from the Sun.

Ask students to tackle the questions on page 379 of the Student Book in pairs. They can then study the Science in context feature on page 379 and make individual notes.

Move on to discuss comets and moons, using the text on page 379 of the Student Book. Students should answer the questions on page 380 in pairs. Discuss the text on satellites on page 381 of the Student Book and the Science in context feature about Satnav and GPS.

Students should then write a sentence on each of the following: planets, moons, comets and satellites. Take feedback and address any misconceptions.

Now discuss orbital speed. Students can tackle the worked example on page 382 of the Student Book without the given solution. After a given period when students have worked out their solutions, their solutions can be compared with the given solution and differences discussed. Students should then tackle the questions on page 382 of the Student Book with a partner.

Bring the activity together

Return to the spider diagram that you created at the beginning of the lesson. Amend it as necessary with the help of the students.

Answers

Page 379

1. Mercury, Venus, Earth, Mars, Jupiter, Saturn, Uranus, Neptune.

2. Inner four = solid, outer four = gaseous.

3. A planet is an object that orbits the Sun and is large enough to have become round due to the force of its own gravity. A planet also has to dominate the neighbourhood around its orbit.

4. Pluto does not dominate its neighbourhood. It is only about double the size of its large 'moon' Charon, but all the true planets are far larger than their moons. Also, planets that dominate their neighbourhoods 'sweep up' asteroids, comets, and other debris, clearing a path along their orbits. Pluto does not.

Page 380

1. Diagram of Solar System with central Sun, a planet in circular orbit around it, a moon in circular orbit around the planet and a comet in highly elliptical orbit around the Sun.

2. Moons are objects that orbit planets rather than the Sun, as planets and comets do.

3. Scientists think that a small planet collided with the Earth soon after it was formed 4.5 billion years ago, forming our Moon.

Page 382

1. a) 28 000 km/h.

   b) 7795 m/s.

2. 7900 m/s.

3. 1021 m/s.

Activity P8b.2 Gravitational field strength

Specification reference: P8b.8.3, P8b.8.4

Learning objectives

- Understand why gravitational field strength, *g*, varies and know that it is different on other planets and the Moon from that on the Earth
- Explain that gravitational force: causes moons to orbit planets, causes the planets to orbit the Sun, causes artificial satellites to orbit the Earth and causes comets to orbit the Sun

Learning outcomes

- State that gravity is a force that affects objects with mass but which is different in other places in the Universe
- Describe how gravity is the force that causes objects in the Universe to follow orbital paths

Resources

Student Book page 383

Resources for class practical and demonstrations (see Technician's notes, following)

Approach

Introduce the activity

Discuss the nature and origin of the force between two magnets (show this using two magnets) and then between two charged balloons (use two balloons each tied to a string and charge them using a duster). Discuss whether there is a force between two people close together. If so, is this the same as that between the Sun and the planets?

Develop the activity

Ask the question 'What is the relationship between the mass of an object and its weight on the Earth's surface?' Students should try to answer this question in small groups using force (newton) meters and a range of known masses. Discuss their suggestions. The final conclusion should be that a 1 kg mass has a weight of about 10 N.

Whirl a rubber bung attached to a length of string in a horizontal circle to introduce the idea of a force towards the centre of the circle. Note that you should wear eye protection while you do this and that you need quite a bit of space – outside would be ideal.

Students observe the effect of increasing the speed of rotation. Link this to the behaviour of planets in the Solar System. Discuss the size of the gravitational force on a planet orbiting the Sun in a circular orbit and the speed of the planet. How would the force and speed of the planet change if the orbit were elliptical?

Discuss 'Gravitational field strength' on page 383 of the Student Book before the students tackle the worksheet individually. They can then tackle the questions on page 383 with a partner.

Move on to discuss the Universe beyond the Solar System. Explain that different models of the Universe's size and shape have developed over time. Explain the current theory which is that the Universe began as a tiny dot which started to expand rapidly about 15 billion years ago and that it is still expanding. Illustrate this with a balloon with dots drawn on the skin to illustrate galaxies.

Bring the activity together

Ask students to work with a partner to write down two things that they have learnt in this activity and one question that they would still like to be answered. If possible, they should research the answer to their question for homework.

Technician's notes

Be sure to check the latest safety notes on these resources before proceeding.

The following resources are needed for the class demonstrations P8b.2:

Introduction

two balloons, each tied to a string
a duster (to charge the balloons)
two magnets

Be aware that some students may have a real fear of balloons.

Main part of the activity

a rubber bung attached to a length of string
eye protection
balloon with dots drawn on the skin to illustrate galaxies

The following resources are needed for the short class practical, per group:

force (newton) meter
a set of known masses (0.1 kg, 0.2 kg, etc.)

There should be enough masses and newton meters for the students to work either in small groups or pairs.

Answers

Page 383

1. 96 N on the Moon, 1560 N on Jupiter.

2. They weigh <u>less</u> on the Moon, but there is gravity there (or else they would float off into space if truly weightless).

3. 546 N.

Activity P8b.3 Consolidation and summary

Learning objectives

- To review the learning points of the topic
- To test understanding through answering questions

Learning outcomes

- Be familiar with the knowledge and understanding summarised in the End of topic checklist
- Be able to apply this knowledge and understanding by answering End of topic questions

Resources

Student Book pages 384–385

Approach

Introduce the activity

Ask students to work with a partner to make a list of key words from this topic. They could then work together to produce a spider diagram showing how the different concepts are linked. They could compare their list with the list of key words given on page 384 of the Student Book.

Discuss the checklist given on page 384 and use questioning to see how much of the content students are comfortable with.

Develop the lesson

Ask students to work individually to work through the End of topic questions on page 385 of the Student Book without reference to the text. As they work, walk around the classroom observing their answers and questioning them as necessary to find out which questions are causing difficulties.

Bring the lesson together

After a set period, ask the students to stop working and discuss any areas of difficulty you observed as you walked round the class.

Students should complete any unanswered questions for homework, but you should stress that they should attempt the questions without reference to the text to see how much content they have retained.

Answers

End of topic questions mark scheme

Question	Correct answer	Marks
1 a)	nearly circular	1
1 b)	it is elliptical	1
2	the speed of a comet is relatively low when further away from the Sun because the force of gravity acting on the comet is very small the speed of a comet is relatively high when close to the Sun because the force of gravity acting on the comet is higher	1 1 1 1
3 a)	any object that is in orbit around another	1
3 b)	the force of gravity pulls the satellite so that it is always accelerating toward the centre of the Earth this force pulls the satellite into a curved orbit if the satellite moved slower the gravitational force would be too great and it would fall to Earth if the satellite moved faster the gravitational force would not be high enough to keep it in a stable orbit and it would fly off into space	1 1 1 1 1 1 1
4 a)	3053 m/s	1
4 b)	5450 m/s	1
5 a)	weight depends on mass and gravitational field strength mass of an object remains the same on Earth and on the Moon but gravitational field strength on the Moon is much less than that on Earth so object will weigh less on the Moon	1 1 1
5 b)	112 N	1
6	a moon orbits a planet whereas planets orbit a star	1 1
	Total:	22

P8c Stellar evolution

Introduction

Stellar evolution explores how stars are classified and how this is related to their temperature and colour. Ways of measuring stars, such as the absolute and apparent magnitude, are studied and use is made of the Hertzsprung–Russell diagram (HR diagram). The life cycles of stars are explained from their origins as huge gas clouds to their ultimate state of a white dwarf or black hole.

Links to other topics

Section	Essential background knowledge	Useful links
7 Radioactivity and particles	P7c Fission and fusion	http://www.bbc.co.uk/schools/gcsebitesize/science/add_aqa_pre_2011/radiation/nuclearfissionrev1.shtml

Topic overview

P8c.1	**The colour of stars** In this activity students learn the classification of stars by their colour and surface temperature. They learn about the luminosity, apparent and absolute magnitude of stars and apply these to the Hertzsprung–Russell diagram (HR diagram).
P8c.2	**Life cycle of stars** In this activity students research the formation of a star. Further work develops understanding of the life cycle of stars with different mass and exposes the physics that explains the stages and their transitions.
P8c.3	**Consolidation and summary** This activity provides an opportunity for a quick recap on the ideas encountered in the topic as well as time for the students to answer the End of topic questions in the Student Book.

Activity P8c.1 The colour of stars

Specification reference: P8c.8.7, P8c.8.8, P8c.8.11P, P8c.8.12P

Learning objectives

- Understand how stars can be classified according to their colour
- Know that a star's colour is related to its surface temperature
- Understand how the brightness of a star at a standard distance can be represented using absolute magnitude
- Draw the main components of the Hertzsprung–Russell diagram (HR diagram)

Learning outcomes

- Be able to classify stars by their colour (which is related to their surface temperature)
- Be able to describe the evolution of stars similar to the Sun.
- Understand the Hertzsprung–Russell diagram

Common misconceptions

Most children when asked to draw a star will draw it as a bright **white** object. Students and adults alike also mainly assume that stars are white. The initial observations of the night sky suggested in the activity may get students to question this belief and open their minds to the existence and explanations for different coloured stars. Also, stars in the sky do not appear to be different distances from the Earth yet they are. The ideas of apparent and absolute magnitude will highlight and hopefully dismiss this often-held misconception.

Resources

Student Book pages 387, 390

Access to night sky beforehand

Access to a Smartboard or equivalent

Access to PC / laptop and Internet

Bunsen burner, tongs, eye protection, piece of steel to heat, heatproof mat

Approach

Introduce the activity

Before the lessons start, a useful exercise is to get students to observe the night sky (with naked eye, binoculars or telescope). They could be set this as homework a week or so beforehand. This would increase the chances of them finding a cloudless sky. They can look to estimate how many stars they see, their colours, their apparent sizes, etc. Some may be able to identify some stars, constellations or planets. Their findings can be shared with the group as a lesson starter.

Questions can be posed to the students such as: 'How long have the stars been there?', 'How are stars formed?', 'Why are some stars different colours?' The responses to this sort of question (and others the students ask) can be stored or displayed on post-it notes for future lessons.

Develop the activity

They can then be introduced to the topic from the Student Book on coloured stars (page 387). A simple demonstration using a Bunsen burner to heat steel will show the different colours (e.g. red hot, orange hot, etc.) depending on how long it is heated or which part of the flame is used. Ask the students to suggest how the colours are related to temperature. Also, you could show a video clip of welding or lightning where temperatures approaching white hot may be reached.

Bring the activity together

The students can use the Student Book to look at the types of star shown in the table on page 387. The scale of temperatures, colours and types of star are summarised here.

This is a good time to ask students to suggest how scientists find out the temperatures of stars. The ideas of luminosity, absolute and apparent magnitude can be explained here too. These can then be related to the Hertzsprung–Russell diagram (or HR diagram).

They can then do some web-based research to see if they can find examples of real stars, their names, type, distance from Earth, surface temperature and colour. These can be presented in a spreadsheet.

Answers

Page 387

1 a) The colour of light that is emitted depends on the surface temperature of the star. Red stars are much cooler than blue stars. These stars emit lower frequency red light.

b)

Type of star	Colour	Approximate surface temperature (K)
O	blue	> 25 000
B	blue	11 000–25 000
A	blue	7500–11 000
F	white to blue	6000–7500
G	yellow to white	5000–6000
K	red to orange	3500–5000
M	red	< 3500

Page 390

1. Radiation cannot escape from a black hole because of the extremely strong gravitational field. Neutron stars are non-luminous.

2. a) Top left of the HR diagram.

   b) These stars will become hotter and brighter and probably explode as a supernova.

   c) They will move towards the top left of the HR diagram.

3. It is in its main sequence phase now. After this it will cool and become a red giant. It will move from the middle lower part of the HR diagram upwards to the right.

Activity P8c.2 Life cycle of stars

Specification reference: P8c.8.9, P8c.8.10

Learning objectives

- Describe the evolution of stars of similar mass to the Sun through the following stages: nebula, star (main sequence), red giant, white dwarf
- Describe the evolution of stars with a mass larger than the Sun

Learning outcomes

- Be able to produce a flow chart or diagram to show the life cycle for stars like our Sun, heavyweight stars and superheavyweight stars

Common misconceptions

This is a topic that is rich in vocabulary and that vocabulary can be confusing to students. Some of the vocabulary that is around is fictitious (and admittedly tempting) such as 'Red Dwarf'. Should students finding themselves writing about Red Dwarf then it shouldn't be in a physics test or exam. It is worthwhile giving key words to students and they can make a glossary of the terms as they arise in their study of the topic. The term 'life cycle' of a star is referred to frequently. It is worth stressing that stars are **not** alive and that the phrase is merely an analogous term.

Many students have a tentative idea of the Big Bang and often apply this incorrectly and simplistically to the formation of planets and stars. This misconception is also one to be aware of.

Resources

Student Book pages 388–389

Access to a Smartboard or equivalent

Access to PC / laptop and Internet

Approach

Introduce the activity

Having learned how to categorise stars in a previous activity, the students can then explore how they are formed. Ask the students if they have any ideas as to how the Sun was formed. Also, you could ask how scientists think they know about stars and how they collect evidence. These ideas from the class can be gathered, shared and tested by the research that follows.

Develop the activity

This can be a research web-based activity. However, it is sensible to restrict this to how stars form – i.e. the nebula phase up to fusion and main sequence. After that it becomes diverse and confusions can occur as stars behave differently depending on their mass. The ideas and the stages can then be built up from the basics the students have discovered. These ideas can be used as a basis for building up more formally the stages of the life cycle of a star.

It is probably better to look at the life cycle of a star like our Sun. The stages can be built up with explanations of each stage, referring to gravity, high temperatures and pressure, nuclear fusion etc. Students can draw a diagram or flow chart showing the stages and annotate it with the explanations for each stage or change in stage.

Bring the activity together

Here is a good opportunity to discuss stars that are much more massive than our Sun. A useful story of the life cycle of different stars can be found on the NASA or STEM websites. For example, this five-minute video clip could be used: https://www.stem.org.uk/elibrary/resource/29954

Using the Student Book, video clips, further web-based research and other resources that are available, students could produce another diagram or flow chart showing the stages of larger stars and annotate it with the explanations for each stage or change in stage. They can then be introduced to the idea of neutron stars, supernovae and black holes. There are numerous photographs and graphics about these on the NASA website. This is a useful time for students to summarise their understanding of this topic . They can produce a presentation which covers the life cycle and stages of stars with different masses.

Activity P8c.3 Consolidation and summary

Learning objectives

- To review the learning points of the topic
- To test understanding through answering questions

Learning outcomes

- Be familiar with the knowledge and understanding summarised in the End of topic checklist
- Be able to apply this knowledge and understanding by answering End of topic questions

Resources

Student Book pages 391–392

Approach

Introduce the activity

Ask students to work with a partner to make a list of key words from this topic. They could then work together to produce a spider diagram showing how the different concepts are linked. They could compare their list with the list of key words given on page 391 of the Student Book.

Discuss the checklist given on page 391 and use questioning to see how much of the content students are comfortable with.

Develop the lesson

At the next opportunity they can work individually through the End of topic questions on page 392 of the Student Book without reference to the text. As they work, walk around the classroom observing their answers and questioning them as necessary to find out which questions are causing difficulties.

Bring the lesson together

After a set period, ask the students to stop working and discuss any areas of difficulty you observed as you walked round the class.

Students should complete any unanswered questions for homework, but you should stress that they should attempt the questions without reference to the text to see how much content they have retained.

Answers

End of topic questions mark scheme

Question	Correct answer	Marks
1	gas particles in a large cloud (nebula) are pulled inwards by gravitational attraction to concentrate the matter together as a proto-star	1
	the temperature increases as the gas becomes more compressed	1
	very high temperature and pressures causes hydrogen nuclei to combine in a fusion reaction	1
	the star is then in its main sequence phase and produces energy steadily for a very long time	1
	when the hydrogen starts to run out the star expands to become a red giant	1
	it then cools and shrinks and ends its life as a white dwarf	1
2	a star with a higher mass starts its life from a larger nebula	1
	as before, gas particles are pulled in by gravitational attraction to concentrate the matter together as a proto-star	1
	the temperature increases as the gas becomes more compressed	1
	and very high temperature and pressures causes hydrogen nuclei to combine in a fusion reaction	1
	the star is then in its main sequence phase and produces energy steadily until the hydrogen starts to run out when fusion of helium to other elements begins	1
	the star gets hotter and hotter	1
	when the helium runs out and fusion stops the star is unstable	1
	and just like a low-mass star it expands	1
	it becomes a red supergiant	1
	eventually stars this massive explodes as a supernova	1
3	black holes have a very large mass, relatively small size and hence a very strong gravitational field	1
	the gravitational field of a black hole is so strong that it warps the space around into such a tight curve that all paths lead into the black hole	1
	therefore, no radiation escapes the black hole's gravitational field	1

Question	Correct answer	Marks
4	nuclear fusion can occur when two hydrogen nuclei are forced together	1
	to produce a larger helium nucleus, a neutron and a large amount of energy	1
	this energy comes from the small amount of mass that is lost in the reaction	1
5	an HR diagram (or Hertzsprung–Russell diagram) for a group of stars is one that shows the luminosity of each star plotted against its surface temperature	1
	it is useful as it shows what stage a star is at in its life cycle	1
	the star's position on the diagram can be used to predict the life cycle of the star	1
6	the apparent magnitude is how bright an object is when seen by an observer on Earth	1
7	the absolute magnitude is the apparent magnitude of an object if it was a standard distance from the Earth	1
	in this way, we can compare many stars regardless of how distant they are	1
	Total:	28

P8d Cosmology

Introduction

The nature of stars has fascinated human beings for generations. It has been an ongoing puzzle just waiting for amateurs and scientists to solve. Despite being some way into the age of space travel we are still unsure about some of the evidence we have. However, we do have a clearer idea about stars, their formation and subsequent life cycle. A bigger picture though is how the Universe began and the Big Bang has dominated our thinking for some time.

Links to other topics

Section	Essential background knowledge	Useful links
1 Forces and motion	P1c Forces, movement, shape and momentum	http://www.bbc.co.uk/schools/gcsebitesize/science/ocr_gateway/energy_resources/big_bangrev1.shtml
3 Waves	P3b Properties of waves P3c The electromagnetic spectrum	http://www.bbc.co.uk/schools/gcsebitesize/science/aqa/origins/redshiftrev2.shtml

Topic overview

P8d.1	**The Big Bang** In this activity students learn to name and describe objects in the Solar System – planets, comets, moons and satellites.
P8d.2	**Consolidation and summary** This activity provides an opportunity for a quick recap on the ideas encountered in the topic as well as time for the students to answer the End of topic questions in the Student Book.

Activity P8d.1 The Big Bang

Specification reference: P8d.8.13, P8d.8.14, P8d.8.15, P8d.8.16, P8d.8.17, P8c.8.18

Learning objectives

- Describe the past evolution of the Universe and the main arguments in favour of the Big Bang
- Describe evidence that supports the Big Bang theory (red-shift and cosmic microwave background (CMB) radiation)
- Describe that if a wave source is moving relative to an observer there will be a change in the observed frequency and wavelength
- Use the equation relating change in wavelength, wavelength, velocity of a galaxy and the speed of light: $\frac{\text{change in wavelength}}{\text{reference wavelength}} = \frac{\text{velocity of a galaxy}}{\text{speed of light}}$

$$\frac{\lambda - \lambda_0}{\lambda_0} = \frac{\Delta\lambda}{\lambda_0} = \frac{v}{c}$$

- Describe the red-shift in light received from galaxies at different distances away from the Earth
- Explain why the red-shift of galaxies provides evidence for the expansion of the Universe

Learning outcomes

- Appreciate how CMB and red-shift are linked to the Big Bang theory
- Be able to explain the relative red-shift of nearby and distant galaxies and use this as a basic explanation of the timing of the Big Bang.
- Be able to do calculations using the idea of red-shift

Common misconceptions

Many students think the Big Bang theory explains what **caused** the original rapid expansion (or an 'explosion' as it is often simplistically thought of). It attempts to explain the observations and evidence there is available – such as red-shift and CMB.

Other issues, which are not misconceptions, may be based around belief and religion. Teachers must be aware of these and make sensible and sensitive judgements about the extent to which these must be considered and respected.

Resources

Student Book pages 394–398

Access to a Smartboard or equivalent

Access to PC / laptop and Internet

Approach

Introduce the activity

Ask the students to work in groups and summarise a few theories they have heard of about how the Universe started. Discuss these ideas and share them with the rest of the class. They can be displayed in post-it form and referred to later.

This is a useful time to revisit the electromagnetic spectrum (in terms of frequency and wavelengths) and the Doppler effect (in terms of sound and waves). Perhaps some of the activities from previous topics could be used here (e.g. The Doppler Ball).

Develop the activity

The students can then be shown spectra from stars showing red-shift. This can then be explained in terms of the Doppler effect. There are numerous websites that show video, graphics and animations which illustrate the Doppler effect.

- This can be further developed to show what scientists have found out from red-shift:

  (the great majority of) galaxies show red-shift which shows they are moving away at great speed
- Distant galaxies show more red-shift than nearby ones which shows that distant galaxies are moving away even quicker than nearby ones
- Extrapolating back in space and time can give scientists a few clues about the time and place of the Big Bang. The ideas can also be extended to show how scientists do calculations involving red-shift in order to calculate the speeds of galaxies. There is a worked example in the Student Book and students would benefit from practising other calculations at this stage

Further to this, there is the incidence of cosmic microwave background radiation which is evenly received from all parts of the Universe.

There are many video clips and animations that show the Big Bang and are helpful in developing the students' understanding.

Bring the activity together

The students can do some web-based research about the Big Bang. They can find out also about red-shift and CMB. They can present their findings in terms of a presentation that explains what the Big Bang is (and was). Also they can look at what may happen in the future. These presentations could usefully be populated with web-links to useful clips and animations. In that way they are then useful for revision and sharing with others.

Answers

Page 395

1. a) The Doppler effect explains that wavelength or frequency of waves changes if the source of the waves is moving relative to the observer. The wavelength of light emitted by a galaxy that moves towards us at speed will decrease. This means it shifts towards the blue end of the visible spectrum, i.e. has a higher frequency.

   b) Red-shifted galaxies are moving away from us. Blue-shifted galaxies are moving towards us and could collide with our galaxy at some time in the future.

Page 396

1. change in wavelength: $\Delta\lambda = 760 - 680 = 80$

$$\frac{\Delta\lambda}{\lambda_0} = \frac{v}{c}$$

$$v = \frac{80 \times 3 \times 10^8}{680}$$

Velocity of the quasar is 3.5×10^7 m/s (to 2 s.f.)

The light is red-shifted so the direction of movement is away from the Earth.

Activity P8d.2 Consolidation and summary

Learning objectives

- To review the learning points of the topic
- To test understanding through answering questions

Learning outcomes

- Be familiar with the knowledge and understanding summarised in the End of topic checklist
- Be able to apply this knowledge and understanding by answering End of topic questions

Resources

Student Book pages 399–400

Approach

Introduce the activity

Ask students to work with a partner to make a list of key words from this topic. They could then work together to produce a spider diagram showing how the different concepts are linked. They could compare their list with the list of key words given on page 399 of the Student Book.

Discuss the checklist given on page 399 and use questioning to see how much of the content students are comfortable with.

Develop the lesson

Ask students to work individually to work through the End of topic questions on page 400 of the Student Book without reference to the text. As they work, walk around the classroom observing their answers and questioning them as necessary to find out which questions are causing difficulties.

Bring the lesson together

After a set period, ask the students to stop working and discuss any areas of difficulty you observed as you walked round the class.

Students should complete any unanswered questions for homework, but you should stress that they should attempt the questions without reference to the text to see how much content they have retained.

Answers

End of topic questions mark scheme

Question	Correct answer	Marks
1	red-shift occurs when a light source moves away from an observer	1
	the wavelength of the radiation that is emitted from it increases because of the Doppler effect	1
	the light from the object appears more red because longer wavelengths of light are at the red end of the visible light spectrum	1
	blue-shift occurs when a light source moves towards an observer	1
	these objects appear more blue because shorter wavelengths of light are at the violet end of the visible light spectrum	1
2 a)	CMB is detected on Earth by radio telescopes	1
	it can be detected in space by some satellites such as the COBE satellite telescope	1
2 b)	In the 1960s microwave radiation	1
	from space was detected by radio telescopes	1
	CMB was uniform / it came from all directions of the Universe	1
	scientists also measured the temperature of this radiation to be 2.7 K	1
	space telescopes have since confirmed the findings	1
3 a)	the galaxies all show red-shift, so are all galaxies are moving away from the Earth / away from each other	1
	the differences in the size of the red shift imply that galaxies are moving away at different speeds	1
	the more red-shifted galaxies, which are more distant, are moving faster than the less red-shifted galaxies, which are nearer	1
3 b)	distant and nearby galaxies all moving away from the same point in space	1

Section 8: Answers to exam-style questions

Question	Correct answer	Marks
2 a) i)	A	1
2 a) ii)	D	1
2 b)	88 days = 7.6×10^6 s; 58 million km = 5.8×10^{10} m orbital speed = $2 \times \pi \times 5.8 \times 10^{10}/7.6 \times 10^6$ = 47 950 m/s or 4.795×10^4 m/s	1 1 1
2 c)	a galaxy is a collection of billions of stars the Universe is a collection of billions of galaxies	1 1
3 a)	emits red light and has a surface temperature of less than 3500 K as it is very large it is probably a red giant	1 1 1
3 b)	red giant will become unstable when fusion ends collapses to become a white dwarf	1 1
	Total:	28

Activity number and name	Learning objectives and outcomes	Specification reference	Student Book pages	Worksheet/ Practical
P1b.1 Distance–time graphs, speed and velocity	**Learning objectives** • Use measurement systems and the appropriate units • Describe motion in a graphical form by plotting and explaining distance–time graphs • Use the relationship between average speed, distance moved and time taken: $\text{average speed} = \frac{\text{distance moved}}{\text{time taken}}$ **Learning outcomes** • Explain the importance of having a consistent measuring system • Draw graphs of distance against time for data given in different forms • Plot and explain distance–time graphs (approach, recession, stationary) • Be able to explain the difference between average speed and speed at a particular moment • Describe average speed as the distance travelled in a set time • Be able to investigate the motion of everyday objects, such as toy cars or tennis balls	1.3–1.5	10–18	Measuring Distance–time data Distance–time graphs

Activity number and name	Learning objectives and outcomes	Specification reference	Student Book pages	Worksheet/ Practical
P1b.2 Acceleration and velocity–time graphs	**Learning objectives** • Introduce velocity as a vector quantity • Define acceleration as rate of change of velocity • Use the relationship between acceleration, change in velocity and time taken: $\text{acceleration} = \frac{\text{change in velocity}}{\text{time taken}}$, $a = \frac{(v-u)}{t}$ • Plot and explain velocity–time graphs, determining acceleration from the gradient and distance travelled from the area between the graph and the time axis • Carry out calculations using the equation $v^2 = u^2 + (2 \times a \times s)$ **Learning outcomes** • Know the definition of acceleration • Be able to calculate acceleration given the relevant data • Describe how to measure the velocity in simple situations (e.g. toy cars accelerating across a table top) • Be able to draw a graph of velocity against time, given relevant data • Be able to use $\text{acceleration} = \frac{\text{change in velocity}}{\text{time taken}}$ to calculate acceleration • Be able to determine acceleration (gradient) and distance travelled (area under graph) from a velocity–time graph • Be able to calculate one of final speed, initial speed, acceleration and distance moved, using the equation $v^2 = u^2 + (2 \times a \times s)$	1.6–1.10	19–25	Measuring acceleration Acceleration calculations
P1b.3 Consolidation and summary	**Learning objectives** • To review the learning points of the topic • To test understanding through answering questions **Learning outcomes** • Be familiar with the knowledge and understanding summarised in the End of topic checklist • Be able to apply this knowledge and understanding by answering End of topic questions	1.3–1.10	26–28	

Activity number and name	Learning objectives and outcomes	Specification reference	Student Book pages	Worksheet/ Practical
P1c.1 Introducing forces	**Learning objectives** • Describe typical effects of forces – changing the speed, shape or direction of objects • Identify different types of force • Understand that force is a vector quantity • Calculate the resultant force in simple situations (one dimension only) • Know that friction is a force that opposes motion **Learning outcomes** • Be able to name at least two different types of force • Be able to draw and interpret diagrams involving multiple forces in familiar situations	1.11–1.16	29–34	Finding forces Multiple forces
P1c.2 Balanced and unbalanced forces	**Learning objectives** • Investigate balanced and unbalanced forces **Learning outcomes** • Know what balanced and unbalanced forces are	–	35–36	Balanced forces
P1c.3 Force, mass and acceleration	**Learning objectives** • Know and use the relationship between force, mass and acceleration: force = mass × acceleration $F = ma$ **Learning outcomes** • Understand the link between force, mass and acceleration • Carry out calculations using $F = ma$ • Apply the principles involved in a range of familiar and unfamiliar situations	1.17	36–38	Force and acceleration
P1c.4 Mass and weight	**Learning objectives** • Know and use the relationship between weight, mass and gravitational field strength: weight = mass × gravitational field strength $W = mg$ **Learning outcomes** • Know how to calculate the weight of an object using $W = mg$	1.18	38–41	

Activity number and name	Learning objectives and outcomes	Specification reference	Student Book pages	Worksheet/ Practical
P1c.5 Falling objects	**Learning objectives** • Describe the forces acting on falling objects • Explain why falling objects react a terminal velocity **Learning outcomes** • State that the acceleration of free fall for a body near to the Earth is constant • Describe qualitatively the motion of bodies falling in a uniform gravitational field with and without air resistance (including reference to terminal velocity)	1.21	41–42	Falling cupcake cases 1 Falling cupcake cases 2
P1c.6 Factors affecting stopping distances	**Learning objectives** • Know that the stopping distance of a vehicle is made up of the sum of the thinking distance and the braking distance • Describe the factors affecting stopping distance (speed, mass, reaction time and road conditions) **Learning outcomes** • Be able to describe how the stopping of a vehicle is related to factors involving the driver, the vehicle and the external conditions • Understand the terms thinking distance, braking distance and stopping distance • Relate changes in these factors to an increase or a decrease in the affected distance	1.19, 1.20	42–44	Reaction time Braking distance

Activity number and name	Learning objectives and outcomes	Specification reference	Student Book pages	Worksheet/ Practical
P1c.7 Turning effect of a force and stability of an object	**Learning objectives** • Define moment as force × perpendicular distance from pivot to line of action of force • Know that the weight of a body acts through its centre of gravity **(P)** • Introduce the principle of moments and conditions for equilibrium • Balance turning forces in simple situations **Learning outcomes** • Be able to describe the turning effect of a force in simple examples • Be able to state the principle of moments • Describe moments as anticlockwise or clockwise and calculate moments, in Nm, in simple situations • Know what the centre of gravity of a body is • Apply the concept of moments to explain the operation of simple devices, e.g. screwdriver • Calculate missing forces and/or distances in balanced situations	1.30P–1.33P	44–49	Finding the centre of gravity Investigating moments Balancing forces
P1c.8 Stretching	**Learning objectives** • Investigate how extension varies with applied force for springs, wire and rubber bands • Know that the linear region of a force–extension graph is linked with Hooke's law • Describe the elastic behaviour of a spring **Learning outcomes** • Be able to state Hooke's law • Describe an experiment to test a spring to see if it obeys Hooke's law • Be able to state the requirements for a graph to show direct proportionality • Describe how not all materials follow Hooke's law when under tension • Describe an experiment to test a material that does not give proportional behaviour (e.g. a rubber band) including safety precautions	1.22–1.24	49–55	Stretching springs

Activity number and name	Learning objectives and outcomes	Specification reference	Student Book pages	Worksheet/ Practical
P1c.9 Momentum	**Learning objectives** • Introduce the definition of momentum **(P)**: momentum = mass × velocity $p = m \times v$ • Apply the law of conservation of momentum to simple situations **(P)** • Use the principle of conservation of momentum to calculate mass, velocity or momentum of objects **(P)** **Learning outcomes** • Be able to calculate momentum • Describe simple situations where momentum is conserved • Make relevant calculations using $p = m \times v$ in simple situations	1.25P, 1.27P	56	Collisions Explosions
P1c.10 Safety features in vehicles	**Learning objectives** • Use the idea of momentum to explain safety features in vehicles, such as seatbelts, airbags and crumple zones **(P)** • Describe force as rate of change of momentum **(P)**: $\text{force} = \frac{\text{change in momentum}}{\text{time taken}}$ $F = \frac{(mv - mu)}{t}$ **Learning outcomes** • Be able to describe the operation of a number of safety features in vehicles in terms of extended stopping times • Use the idea of rate of change of momentum to relate this to reduced forces	1.26P, 1.28P	56–60	
P1c.11 The third law	**Learning objectives** • Demonstrate an understanding of Newton's third law **(P)** **Learning outcomes** • Describe the conditions for a pair of forces to be a third law pair • Give simple examples of third law pairs	1.29P	61–63	

Activity number and name	Learning objectives and outcomes	Specification reference	Student Book pages	Worksheet/ Practical
P1c.12 Consolidation and summary	**Learning objectives** • To review the learning points of the topic • To test understanding through answering questions **Learning outcomes** • Be familiar with the knowledge and understanding summarised in the End of topic checklist • Be able to apply this knowledge and understanding by answering End of topic questions	1.11–1.33	64–69	
P2b.1 Domestic electrical safety features	**Learning objectives** • Explain the use of a number of safety features (insulation, double insulation, earthing, fuses and circuit breakers) to prevent harm to appliances and users **Learning outcomes** • Explain why safety is an important consideration in the use of electrical appliances • Describe the operation of safety features such as double insulation, fuses, circuit breakers and earth wires • Describe the sequence of how a fuse acts to make a circuit safe	2.2	79–82	Fuse blowing

Activity number and name	Learning objectives and outcomes	Specification reference	Student Book pages	Worksheet/ Practical
P2b.2 Electrical heating and current in a resistor, power, voltage, current and time	**Learning objectives** • Understand why a current in a resistor results in the transfer of electrical energy and an increase in temperature • Describe examples of electrical energy being used to supply a heating effect in a domestic context • Define electrical power and its unit: power = current × voltage $P = I \times V$ • Use the relationship between energy transferred, current, voltage and time: energy transferred = current × voltage × time $E = I \times V \times t$ **Learning outcomes** • Describe examples of electrical heating • Recall that power is the energy transferred per second and is measured in watts • Be able to recall and use the equation $P = I \times V$ • Calculate the current required for an appliance given its power rating and operating voltage • Choose the correct value of fuse for an appliance • Use the equation $E = I \times V \times t$	2.3–2.5	82–87	Kettle power
P2b.3 a.c. and d.c.	**Learning objectives** • Know the difference between an a.c. supply (mains electricity) and a d.c. supply (a cell or battery) • Explain why a.c. is used for mains supply **Learning outcomes** • Describe the difference between a.c. and d.c. in terms of electron movement • Draw sketch graphs to illustrate the difference over time of the two methods	2.6	87–88	a.c. to d.c.

Activity number and name	Learning objectives and outcomes	Specification reference	Student Book pages	Worksheet/ Practical
P2b.4 Consolidation and summary	**Learning objectives** • To review the learning points of the topic • To test understanding through answering questions **Learning outcomes** • Be familiar with the knowledge and understanding summarised in the End of topic checklist • Be able to apply this knowledge and understanding by answering End of topic questions	2.2–2.6	89–91	
P2c.1 Series and parallel circuits	**Learning objectives** • Describe what happens to the current and voltage in series and parallel circuits • Explain why series or parallel circuits are appropriate in particular circumstances **Learning outcomes** • Recall the symbols used to draw circuit diagrams • Describe the relative merits of series and parallel circuits in simple situations (e.g. lighting circuits)	2.7	92–97	Series circuits Parallel circuits
P2c.2 How changing resistance affects current	**Learning objectives** • Understand how the current in a series circuit depends on the voltage and the components in the circuit • Understand the circuit operation for an ohmic conductor (e.g. resistance wire) • Understand the behaviour of a number of non-ohmic conductors • Describe the variation of resistance of light-dependent resistors with illumination and thermistors with current **Learning outcomes** • Define resistance • Describe experiments to find *V-I* graphs for various components • Recognise non-ohmic graphs • Summarise the behaviour of non-ohmic components • Describe experiments to demonstrate the behaviour of non-ohmic components • Recognise the temperature–resistance graph (thermistor) and intensity–resistance graph (LDR) as characteristic of the component	2.8, 2.11	97–100	Voltage and current in a filament lamp Voltage and current in a thermistor

Activity number and name	Learning objectives and outcomes	Specification reference	Student Book pages	Worksheet/ Practical
P2c.3 Voltage, current and resistance	**Learning objectives** • Describe how current varies with voltage in wires, resistors, lamps and diodes • Describe the effect of changing resistance on the current in a circuit • Learn the relationship between voltage, current and resistance: voltage = current × resistance $V = I \times R$ **Learning outcomes** • Use the equation $V = I \times R$ correctly • Be able to explain the flow of current in metallic conductors	2.9, 2.10, 2.12, 2.13	101–107	Voltage–current graphs
P2c.4 Voltage, energy and charge	**Learning objectives** • Know that current is the rate of flow of charge • Know and use the relationship between charge, current and time charge = current × time $Q = I \times t$ • Know that current in solid conductors is the flow of negatively charged electrons • Know that voltage is the energy transferred per unit charge passed, meaning that the volt is a joule per coulomb • Know and use the relationship between voltage, energy and charge energy transferred = charge × voltage $E = Q \times V$ **Learning outcomes** • Be able to use $Q = I \times t$ • Be able to define the volt • Be able to use $E = Q \times V$ • Apply ideas about conservation of charge to explain observations from measurements of current in simple circuits	2.14–2.21	107–110	Voltage

Activity number and name	Learning objectives and outcomes	Specification reference	Student Book pages	Worksheet/ Practical
P2c.5 Consolidation and summary	**Learning objectives** • To review the learning points of the topic • To test understanding through answering questions **Learning outcomes** • Be familiar with the knowledge and understanding summarised in the End of topic checklist • Be able to apply this knowledge and understanding by answering End of topic questions	2.7–2.21	111–116	
P2d.1 Charging by friction – attraction and repulsion	**Learning objectives** • Identify materials which are electrical conductors and electrical insulators **(P)** • Use the word 'charge' and understand the need for two types **(P)** • Explain electrostatic charge in terms of loss and gain of electrons **(P)** • Explain electrostatic phenomena in terms of movement of electrons **(P)** **Learning outcomes** • Conduct experiments to demonstrate simple electrostatic effects • Describe how to charge an insulator by friction and explain why an insulator is needed • Explain why there must be two types of charge, describing simple experiments which support this idea • Be able to link positive and negative charges correctly to the movement of electrons.	2.22P–2.26P	117–121	Friction charging
P2d.2 Uses and hazards of electrostatic charges	**Learning objectives** • Explain hazards associated with electrostatic charges (e.g. fuelling aircraft) • Discuss devices which make use of electrostatic phenomena (e.g. photocopiers) **Learning outcomes** • Be able to describe simple examples where electrostatic effects pose a potential hazard, e.g. in refuelling • Be able to describe measures taken to reduce the associated risks, e.g. use of appropriate materials • Be able to describe the operation of a device, such as a photocopier or a precipitator in a chimney, that use electrostatic effects	2.27P, 2.28P	122–125	

Activity number and name	Learning objectives and outcomes	Specification reference	Student Book pages	Worksheet/ Practical
P2d.3 Consolidation and summary	**Learning objectives** • To review the learning points of the topic • To test understanding through answering questions **Learning outcomes** • Be familiar with the knowledge and understanding summarised in the End of topic checklist • Be able to apply this knowledge and understanding by answering End of topic questions	2.22P–2.28P	126–128	
P3b.1 Properties of waves	**Learning objectives** • Explain the difference between transverse and longitudinal waves • Define the terms amplitude, wavefront, frequency, wavelength and period of a wave • Understand that waves transfer energy and information without transferring matter **Learning outcomes** • State examples of wave motion, including light, sound, water and mechanical waves, distinguishing them as longitudinal or transverse • Use the terms amplitude, wavefront, wavelength, frequency, period and speed to describe waves, including correctly labelling diagrams	3.2–3.4	137–139	Making and seeing waves Waves sheet
P3b.2 The wave equation	**Learning objectives** • Use the wave equation to link wavelength, frequency and wave speed: wave speed = frequency × wavelength $v = f \times \lambda$ • Use the relationship between frequency and time period: frequency = $\frac{1}{\text{time period}}$ $f = \frac{1}{T}$ **Learning outcomes** • Be able to use the equations $v = f \times \lambda$ and $f = \frac{1}{T}$ in a variety of contexts • Remember that speed = distance/time	3.5–3.7	140–141	

Activity number and name	Learning objectives and outcomes	Specification reference	Student Book pages	Worksheet/ Practical
P3b.3 Reflection and refraction	**Learning objectives** • Explain the change in observed frequency and wavelength when the wave source is moving relative to the observer (the Doppler effect) • Explain reflection and refraction in waves **Learning outcomes** • Be able to relate the change in frequency as a result of the Doppler effect to movement towards and away from the observer • Describe the use of water waves to show the behaviour of wavefronts, reflection at a plane surface and refraction due to change of speed	3.8, 3.9	141–145	
P3b.4 Consolidation and summary	**Learning objectives** • To review the learning points of the topic • To test understanding through answering questions **Learning outcomes** • Be familiar with the knowledge and understanding summarised in the End of topic checklist • Be able to apply this knowledge and understanding by answering End of topic questions	3.2–3.9	146–147	
P3c.1 Electromagnetic spectrum	**Learning objectives** • Know that light is part of the electromagnetic spectrum that includes radio, microwave, infrared, visible, ultraviolet, x-ray and gamma ray radiations • Know that all these waves travel at the same speed in free space • Know the regions of the electromagnetic spectrum in order of increasing frequency (decreasing wavelength), including the colours of the visible spectrum **Learning outcomes** • Be able to list the regions of the electromagnetic spectrum in order from radio waves to gamma radiation	3.10, 3.11	148–154	The colours of light

Activity number and name	Learning objectives and outcomes	Specification reference	Student Book pages	Worksheet/ Practical
P3c.2 Uses and hazards of electromagnetic radiation	**Learning objectives** • Explain in detail uses of electromagnetic radiation • Explain hazards relating to excessive exposure to electromagnetic waves and describe some measures to reduce risks to humans **Learning outcomes** • Describe at least one use of each region of the electromagnetic spectrum • Recall possible harmful effects on humans from relevant regions of the electromagnetic spectrum • Explain methods of reducing the risks to humans from waves of the electromagnetic spectrum	3.12, 3.13	150–154	
P3c.3 Consolidation and summary	**Learning objectives** • To review the learning points of the topic • To test understanding through answering questions **Learning outcomes** • Be familiar with the knowledge and understanding summarised in the End of topic checklist • Be able to apply this knowledge and understanding by answering End of topic questions	3.10–3.13	155–157	
P3d.1 Reflection of light	**Learning objectives** • Know that light waves are transverse waves that can be reflected • Use the law of reflection (the angle of incidence equals the angle of reflection) • Draw ray diagrams to illustrate reflection **Learning outcomes** • Describe how light is reflected from plane surfaces, including using ray diagrams • Be able to construct ray diagrams to show how images are formed in plane mirrors	3.14–3.16	158–162	Reflection Images from rays Making a periscope

Activity number and name	Learning objectives and outcomes	Specification reference	Student Book pages	Worksheet/ Practical
P3d.2 Refraction of light	**Learning objectives** • Know that light waves can be refracted • Recall and use the relationship between refractive index, angle of incidence and angle of refraction: $n = \frac{\sin i}{\sin r}$ • Determine refractive index of a glass block **Learning outcomes** • Describe an experimental demonstration of the refraction of light • Describe an experiment to determine the refractive index of glass, using a glass block	3.14, 3.16–3.19	162–166	Refraction Refractive index
P3d.3 Total internal reflection and critical angle	**Learning objectives** • Describe the role of total internal reflection in transmitting information along optical fibres and in prisms • Introduce total internal reflection and the critical angle • Know and use the relationship between critical angle and refractive index: $\sin c = \frac{1}{n}$ **Learning outcomes** • Give the meaning of critical angle, *c* • Know and use the relationship between critical angle and refractive index, $\sin c = \frac{1}{n}$ • Describe total internal reflection in optical fibres and prisms • Describe the action of optical fibres particularly in medicine and communications technology	3.20–3.22	167–170	Critical angle

Activity number and name	Learning objectives and outcomes	Specification reference	Student Book pages	Worksheet/ Practical
P3d.4 Properties of sound	**Learning objectives** • Describe the longitudinal nature of sound waves and how they can be reflected and refracted • Know the range of frequencies for human hearing (20–20 000 Hz) **(P)** **Learning outcomes** • Describe the production of sound by vibrating sources • Describe the longitudinal nature of sound waves • Show an understanding that a medium is needed to transmit sound waves • Relate the loudness and pitch of sound waves to amplitude and frequency • Describe how the reflection of sound may produce an echo • Describe compression and rarefaction • State that the range of human hearing is typically 20–20 000 Hz	3.23, 3.24P	170–172	Sound sources
P3d.5 Measuring the speed of sound	**Learning objectives** • Understand how an oscilloscope and microphone can be used to display a sound wave **(P)** • Relate the pitch of a sound to the frequency of vibration of the source • Relate the loudness of a sound to the amplitude of vibration **Learning outcomes** • Describe an experiment to measure the speed of sound in air • Describe an experiment using an oscilloscope to determine the frequency of a sound wave • Be able to relate the pitch of a sound to its frequency and the loudness of a sound to its amplitude	3.25P–3.29P	172–176	Measuring the speed of sound

Activity number and name	Learning objectives and outcomes	Specification reference	Student Book pages	Worksheet/ Practical
P3d.6 Consolidation and summary	**Learning objectives** • To review the learning points of the topic • To test understanding through answering questions **Learning outcomes** • Be familiar with the knowledge and understanding summarised in the End of topic checklist • Be able to apply this knowledge and understanding by answering End of topic questions	3.14–3.23, 3.24P–3.29P	177–182	
P4b.1 Energy conservation and transfer	**Learning objectives** • Understand the following energy stores: chemical, kinetic, gravitational, elastic, thermal, magnetic, electrostatic, nuclear • Describe the following energy transfers: mechanical, electrical, by heating, by radiation (light and sound) • Use the principle of conservation of energy **Learning outcomes** • Describe energy transfers involving energy stores • Understand that energy is conserved	4.2, 4.3	189–193	Energy
P4b.2 Efficiency and Sankey diagrams	**Learning objectives** • Know and use the relationship $\text{efficiency} = \frac{\text{useful energy output}}{\text{total energy input}} \times 100\%$ • Describe a variety of everyday and scientific devices and situations, explaining the transfer of input energy in terms of the efficiency equation. • Represent energy transfers by Sankey diagrams **Learning outcomes** • Be able to use the idea of efficiency and apply this to a range of energy transfers • Be able to describe energy transfers using Sankey diagrams	4.4, 4.5	194–198	Energy circus Heater efficiency

Activity number and name	Learning objectives and outcomes	Specification reference	Student Book pages	Worksheet/ Practical
P4b.3 Conduction, convection and radiation	**Learning objectives** • Describe how thermal energy transfer may take place by conduction, convection and radiation • Explain the role of convection in everyday phenomena • Explain how emission and absorption of radiation are related to surface and temperature • Explain how insulation is used to reduce energy transfers from buildings and the human body **Learning outcomes** • Be able to describe mechanisms of thermal energy transfer • Understand how knowledge of energy transfer mechanisms is used to reduce energy transfers in a range of situations	4.6–4.10	198–208	Insulating a house
P4b.4 Consolidation and summary	**Learning objectives** • To review the learning points of the topic • To test understanding through answering questions **Learning outcomes** • Be familiar with the knowledge and understanding summarised in the End of topic checklist • Be able to apply this knowledge and understanding by answering End of topic questions	4.2–4.10	209–211	
P4c.1 Work	**Learning objectives** • Know and use the relationship between work, force and distance moved in the direction of the force: work done = force × distance $W = F \times d$ • Understand that work done is equal to energy transferred **Learning outcomes** • Be able to use the equation work done = force × distance • Know the link between forces and energy transfer	4.11, 4.12	212–214	Doing work

Activity number and name	Learning objectives and outcomes	Specification reference	Student Book pages	Worksheet/ Practical
P4c.2 Gravitational potential energy, kinetic energy and conservation of energy	**Learning objectives** • Know and use the relationship between gravitational potential energy, mass, gravitational field strength and height: gravitational potential energy = mass × gravitational field strength × height GPE = $m \times g \times h$ • Know and use the relationship: kinetic energy $= \frac{1}{2} \times \text{mass} \times \text{speed}^2$ KE $= \frac{1}{2} \times m \times v^2$ • Understand how conservation of energy produces a link between gravitational potential energy, kinetic energy and work **Learning outcomes** • Be able to calculate the energy transferred in number of situations	4.13–4.15	215–218	Up and down Pendulum swing
P4c.3 Power	**Learning objectives** • Describe power as the rate of transfer of energy or the rate of doing work • Use the relationship between power, work done (energy transferred) and time taken: power $= \frac{\text{work done}}{\text{time taken}}$ $P = \frac{W}{t}$ **Learning outcomes** • Be able to relate power to the rate at which energy is transferred	4.16, 4.17	218–223	Personal power Power calculations

Activity number and name	Learning objectives and outcomes	Specification reference	Student Book pages	Worksheet/ Practical
P4c.4 Consolidation and summary	**Learning objectives** • To review the learning points of the topic • To test understanding through answering questions **Learning outcomes** • Be familiar with the knowledge and understanding summarised in the End of topic checklist • Be able to apply this knowledge and understanding by answering End of topic questions	4.11–4.17	224–226	
P4d.1 Energy resources	**Learning objectives** • Describe the energy transfers involved in generating electricity using: wind, water, geothermal resources, solar heating systems, solar cells, fossil fuels, nuclear power **(P)** • Describe the advantages and disadvantages of methods of large-scale electricity production from various renewable and non-renewable resources **(P)** **Learning outcomes** • Understand the difference between energy sources and energy resources • Be able to describe advantages and disadvantages of a range of energy resources, giving suitable examples	4.18P, 4.19P	227–234	Investigating a model wind turbine
P4d.2 Consolidation and summary	**Learning objectives** • To review the learning points of the topic • To test understanding through answering questions **Learning outcomes** • Be familiar with the knowledge and understanding summarised in the End of topic checklist • Be able to apply this knowledge and understanding by answering End of topic questions	4.18P, 4.19P	235–237	

Activity number and name	Learning objectives and outcomes	Specification reference	Student Book pages	Worksheet/ Practical
P5b.1 Density, mass and volume	**Learning objectives** • Know and use the relationship between density, mass and volume: $\text{density} = \frac{\text{mass}}{\text{volume}}$ $\rho = \frac{m}{V}$ • Describe experiments to determine density using direct measurements of mass and volume **Learning outcomes** • Be able to calculate densities • Be able to use ideas about density in a variety of situations, for example, to describe why convection currents occur	5.3, 5.4	247–253	Finding the density of an irregularly shaped object
P5b.2 Pressure, force and area	**Learning objectives** • Know and use the relationship between pressure, force and area: $\text{pressure} = \frac{\text{force}}{\text{area}}$ $p = \frac{F}{A}$ **Learning outcomes** • Be able to calculate pressures	5.5	254–256	
P5b.3 Pressure in gases and liquids	**Learning objectives** • Understand how the pressure at a point in a gas or liquid which is at rest acts equally in all directions • Know and use the relationship for pressure difference: pressure difference = height × density × g $p = h \times \rho \times g$ **Learning outcomes** • Be able to use ideas about pressure in a variety of situations, for example, to explain why pressure increases as you go towards the bottom of the ocean • Be able to calculate pressure differences	5.6, 5.7	256–261	

Activity number and name	Learning objectives and outcomes	Specification reference	Student Book pages	Worksheet/ Practical
P5b.4 Consolidation and summary	**Learning objectives** • To review the learning points of the topic • To test understanding through answering questions **Learning outcomes** • Be familiar with the knowledge and understanding summarised in the End of topic checklist • Be able to apply this knowledge and understanding by answering End of topic questions	5.3–5.7	262–264	
P5c.1 Arrangement of particles in solids, liquids and gases	**Learning objectives** • Describe the arrangement and motion of particles in solids, liquids and gases **Learning outcomes** • Know how particles are arranged in solids, liquids and gases	5.10P	266–267	Kinetic theory
P5c.2 Melting and evaporation	**Learning objectives** • Understand the changes that occur when a solid melts to form a liquid, and when a liquid evaporates or boils to form a gas • Explain why heating a system will change the energy stored within the system, raising its temperature or changing its state • Obtain a temperature–time graph to show the constant temperature during a change of state **Learning outcomes** • Be able to explain the terms: melting, boiling, evaporation and condensation	P5c.5.8P, P5c.5.9P, 5.11P	267–271	Candle wax cooling curve Evaporation and condensation

Activity number and name	Learning objectives and outcomes	Specification reference	Student Book pages	Worksheet/ Practical
P5c.3 Specific heat capacity	**Learning objectives** • Know that specific heat capacity is the energy required to change the temperature of an object by one degree Celsius per kilogram of mass (J/kg °C) • Use the equation: change in thermal energy = mass × specific heat capacity × change in temperature $\Delta Q = m \times c \times \Delta T$ • Investigate the specific heat capacity of some materials **Learning outcomes** • Be able to determine experimental values of shc and compare these to accepted values • Be able to explain the practical reasons that cause the differences between experimental values and accepted values	5.12P–5.14P	271–274	Specific heat capacity calculations
P5c.4 Consolidation and summary	**Learning objectives** • To review the learning points of the topic • To test understanding through answering questions **Learning outcomes** • Be familiar with the knowledge and understanding summarised in the End of topic checklist • Be able to apply this knowledge and understanding by answering End of topic questions	5.8P–5.14P	275–276	
P5d.1 Kinetic theory	**Learning objectives** • Explain how molecules in a gas have a random motion and that they exert a force and hence a pressure on the walls of the container **Learning outcomes** • Be able to use the kinetic theory to describe solids, liquids and gases • Be able to apply the kinetic theory in a range of situations, using it to explain the observed behaviour of gases in particular	5.15	277–279	

Activity number and name	Learning objectives and outcomes	Specification reference	Student Book pages	Worksheet/ Practical
P5d.2 The Kelvin temperature scale	**Learning objectives** • Understand why there is an absolute zero of temperature which is –273 ^{0}C • Describe the Kelvin scale of temperature and be able to convert between the Kelvin and Celsius scales • Understand why an increase in temperature results in an increase in the average speed of gas molecules • Know that the Kelvin temperature of the gas is proportional to the average kinetic energy of its molecules **Learning outcomes** • Be able to use the Kelvin scale of temperature	5.16–5.19	280–281	
P5d.3 Pressure, temperature and volume	**Learning objectives** • Explain the qualitative relationships between pressure and volume at constant temperature and between pressure and Kelvin temperature at constant volume • Use the relationship between the pressure and Kelvin temperature of a fixed mass of gas at constant volume: $\frac{p_1}{T_1} = \frac{p_2}{T_2}$ • Use the relationship between the pressure and volume of a fixed mass of gas at constant temperature: $p_1V_1 = p_2V_2$ **Learning outcomes** • Be able to use equations to link pressure, volume and Kelvin temperature of a fixed amount of gas	5.20–5.22	282–285	
P5d.4 Consolidation and summary	**Learning objectives** • To review the learning points of the topic • To test understanding through answering questions **Learning outcomes** • Be familiar with the knowledge and understanding summarised in the End of topic checklist • Be able to apply this knowledge and understanding by answering End of topic questions	5.15–5.22	286–288	

Activity number and name	Learning objectives and outcomes	Specification reference	Student Book pages	Worksheet/ Practical
P6b.1 Introduction to magnetism	**Learning objectives** • Know that magnets repel and attract other magnets and attract magnetic substances • Describe the properties of magnetically hard and soft materials • Understand the term magnetic field line **Learning outcomes** • Understand that magnets repel and attract other magnets • Understand that magnets attract magnetic substances • Be able to describe the properties of magnetically hard and soft materials	6.2–6.4	295–297	
P6b.2 Magnetic field patterns and induction	**Learning objectives** • Understand magnetic field patterns • Know that magnetism is induced in some materials when they are placed in a magnetic field • Describe how to use two permanent magnets to produce a uniform magnetic field pattern **Learning outcomes** • Know how to plot magnetic fields • Understand that magnetism is induced in some materials when they are placed in a magnetic field • Describe experiments to investigate the magnetic field pattern for a permanent bar magnet and that between two bar magnets • Be able to use the idea of magnetic field lines to explain magnetic effects	6.4–6.7	297–303	Finding magnetic field patterns
P6b.3 Consolidation and summary	**Learning objectives** • To review the learning points of the topic • To test understanding through answering questions **Learning outcomes** • Be familiar with the knowledge and understanding summarised in the End of topic checklist • Be able to apply this knowledge and understanding by answering End of topic questions	6.2–6.7	304–305	

Activity number and name	Learning objectives and outcomes	Specification reference	Student Book pages	Worksheet/ Practical
P6c.1 Electromagnets	**Learning objectives** • Know that an electric current in a conductor produces a magnetic field round it **(P)** • Describe the construction of electromagnets **(P)** • Draw magnetic field patterns for a straight wire, a flat circular coil and a solenoid when each is carrying a current **(P)** **Learning outcomes** • Be able to describe how to make an electromagnet • Be able to sketch the magnetic field patterns for wires and coils of various shapes	6.8, 6.9P, 6.10P	306–310	
P6c.2 Force on a charged particle and current-carrying wire	**Learning objectives** • Know that there is a force on a charged particle when it is moves in a magnetic field as long as its motion is not parallel to the field **(P)** • Understand why a force is exerted on a current-carrying wire in a magnetic field, and how this effect is applied in simple d.c. electric motors and loudspeakers **(P)** • Use the left-hand rule to predict the direction of the resulting force when a wire carries a current perpendicular to a magnetic field **(P)** • Describe how the force on a current-carrying conductor in a magnetic field changes with the magnitude and direction of the field and current **(P)** **Learning outcomes** • Understand that there is a force on a charged particle when it moves in a magnetic field as long as its motion is not parallel to the field • Be able to explain how simple d.c. motors and loudspeakers work • Know how to predict the direction of the force on a current-carrying wire perpendicular to a magnetic field • Know how to increase the force on a current-carrying conductor	6.11P, 6.12–6.14	311–316	Building an electric motor

Activity number and name	Learning objectives and outcomes	Specification reference	Student Book pages	Worksheet/ Practical
P6c.3 Consolidation and summary	**Learning objectives** • To review the learning points of the topic • To test understanding through answering questions **Learning outcomes** • Be familiar with the knowledge and understanding summarised in the End of topic checklist • Be able to apply this knowledge and understanding by answering End of topic questions	6.8, 6.9P–6.11P, 6.12–6.14	317–319	
P6d.1 Introduction to electromagnetic induction	**Learning objectives** • Know that a voltage is induced in a conductor or a coil when it moves through a magnetic field or when a magnetic field changes through it and describe the factors which affect the size of the induced voltage • Describe the generation of electricity by the rotation of a magnet within a coil of wire and of a coil within a magnetic field and describe the factors which affect the size of the induced voltage **Learning outcomes** • Be able to describe how electromagnetic induction is used to generate electric currents	6.15, 6.16	320–324	Electromagnetic induction Induction from a coil

Activity number and name	Learning objectives and outcomes	Specification reference	Student Book pages	Worksheet/ Practical
P6d.2 Transformers	**Learning objectives** • Describe the structure of a transformer, and understand that a transformer changes the size of an alternating voltage by having different numbers of turns on the input and output sides **(P)** • Know and use the relationship between the input (primary) and output (secondary) voltages and the turns ratio for a transformer **(P)**: $\frac{\text{input (primary) voltage}}{\text{output (secondary) voltage}} = \frac{\text{primary turns}}{\text{secondary turns}}$ $\frac{V_p}{V_s} = \frac{n_p}{n_s}$ **Learning outcomes** • Be able to describe the structure and function of a transformer • Be able to carry out calculations relating to transformers	6.17P, 6.19P	324–326	Building a transformer
P6d.3 Using transformers in electricity generation	**Learning objectives** • Explain the use of step-up and step-down transformers in the large-scale generation and transmission of electrical energy **(P)** • Know and use the relationship input power = output power $V_P I_P = V_S I_S$ for 100% efficiency **(P)** **Learning outcomes** • Know how electricity is transmitted and how step-up and step-down transformers are used in the process	6.17P–6.20P	326–329	

Activity number and name	Learning objectives and outcomes	Specification reference	Student Book pages	Worksheet/ Practical
P6d.4 Consolidation and summary	**Learning objectives** • To review the learning points of the topic • To test understanding through answering questions **Learning outcomes** • Be familiar with the knowledge and understanding summarised in the End of topic checklist • Be able to apply this knowledge and understanding by answering End of topic questions	6.15, 6.16, 6.17P–6.20P	330–332	
P7b.1 The structure of an atom	**Learning objectives** • Describe the structure of an atom in terms of protons, neutrons and electrons and use symbols such as ${}^{14}_{6}C$ to describe particular nuclei • Know the terms atomic (proton) number, mass (nucleon) number and isotope **Learning outcomes** • Be able to describe the structure of an atom in terms of protons, neutrons and electrons • Be able to use the terms mass (nucleon) number, atomic (proton) number and isotope	7.2, 7.3	339–341	Structure of atoms
P7b.2 Ionising radiations	**Learning objectives** • Know that alpha (α) particles, beta (β^-) particles and gamma (γ) rays are ionising radiations emitted from unstable nuclei in a random process • Describe the nature of alpha (α) particles, beta (β^-) particles and gamma (γ) rays and recall that they may be distinguished in terms of penetrating power and ability to ionise • Describe the effects on the atomic and mass numbers of a nucleus of the emission of each of the four main types of radiation (alpha, beta, gamma and neutron radiation) • Understand how to balance nuclear equations in terms of mass and charge **Learning outcomes** • Be able to describe the nature and properties of the three major emissions from radioactive materials and know their differing penetrating power and ability to ionise • Be able to describe what radioactivity is and why some atoms are radioactive	7.4–7.8	343–347	Radioactivity

Activity number and name	Learning objectives and outcomes	Specification reference	Student Book pages	Worksheet/ Practical
P7b.3 Background radiation and half-life	**Learning objectives** • Know that ionising radiations can be detected using a photographic film or a Geiger-Müller detector • Explain the sources of background (ionising) radiation from Earth and space • Know that the activity of a radioactive source decreases over a period of time and is measured in becquerels • Know the definition of the term half-life and understand that it is different for different radioactive isotopes • Use the concept of half-life to carry out simple calculations on activity, including graphical methods **Learning outcomes** • Be able to describe how radioactivity can be detected and is measured • Be able to use ideas about half-life to describe how radioactive materials behave	7.9–7.13	348–352	
P7b.4 Uses and dangers of radioactivity	**Learning objectives** • Describe the uses of radioactivity in industry and medicine • Describe the difference between contamination and irradiation • Describe the dangers of ionising radiations, including: radiation can cause mutations in living organisms, radiation can damage cells and tissue, the problems arising in the disposal of radioactive waste. • Describe how the risks associated with ionising radiations can be reduced **Learning outcomes** • Be able to describe some uses and hazards associated with radioactive materials	7.14–7.16	352–356	
P7b.5 Consolidation and summary	**Learning objectives** • To review the learning points of the topic • To test understanding through answering questions **Learning outcomes** • Be familiar with the knowledge and understanding summarised in the End of topic checklist • Be able to apply this knowledge and understanding by answering End of topic questions	7.2–7.16	357–361	

Activity number and name	Learning objectives and outcomes	Specification reference	Student Book pages	Worksheet/ Practical
P7c.1 Nuclear fission	**Learning objectives** • Know that nuclear reactions, including fission, fusion and radioactive decay, can be a source of energy • Understand how a nucleus of U-235 can be split (the process of fission) by collision with a neutron, and that this process releases energy in the form of kinetic energy of the fission products • Know that the fission of U-235 produces two daughter nuclei and a small number of neutrons • Describe how a chain reaction can be set up if the neutrons produced by one fission strike other U- 235 nuclei • Describe the role played by the control rods and moderator when the fission process is used as an energy source to generate electricity • Understand the role of shielding around a nuclear reactor **Learning outcomes** • Be able to describe how our knowledge of the atom has led to nuclear fission as an energy resource	7.17–7.22	362–366	Nuclear fission
P7c.2 Nuclear fusion	**Learning objectives** • Explain the difference between nuclear fusion and nuclear fission • Describe nuclear fusion as the creation of larger nuclei resulting in a loss of mass from smaller nuclei, accompanied by a release of energy • Know that fusion is the energy source for stars • Explain why nuclear fusion does not happen at low temperatures and pressures, due to electrostatic repulsion of protons **Learning outcomes** • Be able to describe the nature of attractive and repulsive forces inside the nucleus • Be able to explain how energy is absorbed and released in a fusion reaction • Know the elements of nuclear fusion in stars • Appreciate the longevity of nuclear fusion in stars	7.23–7.26	367–368	

Activity number and name	Learning objectives and outcomes	Specification reference	Student Book pages	Worksheet/ Practical
P7c.3 Consolidation and summary	**Learning objectives** • To review the learning points of the topic • To test understanding through answering questions **Learning outcomes** • Be familiar with the knowledge and understanding summarised in the End of topic checklist • Be able to apply this knowledge and understanding by answering End of topic questions	7.17–7.26	369–371	
P8b.1 The Solar System	**Learning objectives** • Know that the Universe is a large collection of billions of galaxies, that a galaxy is a large collection of billions of stars and that our solar system is in the Milky Way galaxy • Describe the differences in the orbits of comets, moons and planets • Use the relationship between orbital speed, orbital radius and time period $\text{orbital speed} = \frac{2 \times \pi \times \text{orbital radius}}{\text{time period}}$ $v = \frac{2 \times \pi \times r}{T}$ **Learning outcomes** • Be able to distinguish between planets, moons, comets, artificial satellites and stars • Be able to sketch the different orbits of these objects • Be able to calculate the orbital speed of an object in orbit	8.2, 8.5, 8.6	377–382	

Activity number and name	Learning objectives and outcomes	Specification reference	Student Book pages	Worksheet/ Practical
P8b.2 Gravitational field strength	**Learning objectives** • Understand why gravitational field strength, *g*, varies and know that it is different on other planets and the Moon from that on the Earth • Explain that gravitational force: causes moons to orbit planets, causes the planets to orbit the Sun, causes artificial satellites to orbit the Earth and causes comets to orbit the Sun **Learning outcomes** • State that gravity is a force that affects objects with mass but which is different on other places in the Universe • Describe how gravity is the force that causes objects in the Universe to follow orbital paths	8.3, 8.4	383	
P8b.3 Consolidation and summary	**Learning objectives** • To review the learning points of the topic • To test understanding through answering questions **Learning outcomes** • Be familiar with the knowledge and understanding summarised in the End of topic checklist • Be able to apply this knowledge and understanding by answering End of topic questions	8.2–8.6	384–385	
P8c.1 The colour of stars	**Learning objectives** • Understand how stars can be classified according to their colour • Know that a stars colour is related to its surface temperature • Understand how the brightness of a star at a standard distance can be represented using absolute magnitude **(P)** • draw the main components of the Hertzsprung–Russell diagram (HR diagram) **(P)** **Learning outcomes** • Be able to classify stars by their colour (which is related to their surface temperature) • Be able to describe the evolution of stars similar to the Sun. • Understand the Hertzsprung–Russell diagram	8.7, 8.8, 8.11P, 8.12P	387, 390	

Activity number and name	Learning objectives and outcomes	Specification reference	Student Book pages	Worksheet/ Practical
P8c.2 Life cycle of stars	**Learning objectives** • Describe the evolution of stars of similar mass to the Sun through the following stages: nebula, star (main sequence), red giant, white dwarf • Describe the evolution of stars with a mass larger than the Sun **Learning outcomes** • Be able to produce a flow chart or diagram to show the life cycle for stars like our Sun, heavyweight stars and superheavyweight stars	8.9, 8.10	388–389	
P8c.3 Consolidation and summary	**Learning objectives** • To review the learning points of the topic • To test understanding through answering questions **Learning outcomes** • Be familiar with the knowledge and understanding summarised in the End of topic checklist • Be able to apply this knowledge and understanding by answering End of topic questions	8.7–8.10, 8.11P, 8.12P	391–392	

Activity number and name	Learning objectives and outcomes	Specification reference	Student Book pages	Worksheet/ Practical
P8d.1 The Big Bang	**Learning objectives** • Describe the past evolution of the Universe and the main arguments in favour of the Big Bang • Describe evidence that supports the Big Bang theory (red-shift and cosmic microwave background (CMB) radiation) • Describe that if a wave source is moving relative to an observer there will be a change in the observed frequency and wavelength • Use the equation relating change in wavelength, wavelength, velocity of a galaxy and the speed of light: $\frac{\text{change in wavelength}}{\text{reference wavelength}} = \frac{\text{velocity of a galaxy}}{\text{speed of light}}$ $\frac{\lambda - \lambda_0}{\lambda_0} = \frac{\Delta\lambda}{\lambda_0} = \frac{v}{c}$ • Describe the red-shift in light received from galaxies at different distances away from the Earth • Explain why the red-shift of galaxies provides evidence for the expansion of the Universe **Learning outcomes** • Appreciate how CMB and red-shift are linked to the Big Bang theory • Be able to explain the relative red-shift of nearby and distant galaxies and use this as a basic explanation of the timing of the Big Bang • Be able to do calculations using the idea of red-shift	8.13P–8.18P	394–398	
P8d.2 Consolidation and summary	**Learning objectives** • To review the learning points of the topic • To test understanding through answering questions **Learning outcomes** • Be familiar with the knowledge and understanding summarised in the End of topic checklist • Be able to apply this knowledge and understanding by answering End of topic questions	8.13P–8.18P	399–400	

